I0486289

First published 2016

Copyright © S Bowen 2016

All rights reserved. No part of this publication may be
reproduced, distributed, or transmitted in any form or by
any means, including photocopying, recording, or other
electronic or mechanical methods, without the prior
permission of the copyright owner.

ISBN-13: 9781515310020

ISBN-10: 1515310027

Cover image – Budding yeast cells stained with calcofluor-
white fluorescent stain. Credit: The author.

Biblical Theory in a Molecular Era

The

Leaven

Biblical Theory in a Molecular Era

The Leaven

For

my parents and loved ones
Yvette, Justin, and their children.

Biblical Theory in a Molecular Era

contents

Biblical Theory in a Molecular Era

Preface

In humans, the instinct to survive can surmount all other emotions. Transmissible and genetic diseases may have generated fear in ancient civilisations as they were not understood and often attributed to the work of supernatural forces. Extreme measures were taken to avoid angering these powerful deities. Rituals and sacrifices perhaps granting some form of hope against the inexplicable ravages of natural phenomena. This book explores the precarious relationships between fear and uncertainty, scientific research, and religious belief, in relation to the philosophies in the Old and New Testaments of the Bible.

In analysing the uninformed behaviour of Old Testament psyche with a modern scientific rational, a logical, though seemingly inhumane, answer to disease management is revealed. Gradually, the extreme methods of disease management enforced by regulations in the Old Testament were replaced by a more compassionate Greco-Roman influenced approach in the New Testament that prescribed treatment rather than ostracism.

Scientific advances in microbiology and immunology now largely replace religious methods of disease prevention, but do not completely obliterate fear and uncertainty. Controversies involving the corruptive influences that can lead to fatalities and human misery are discussed at length within this work, as are conflicts between religion and science in the molecular era. The gradual domestication of leaven into yeast is used to organise and focus the text.

Yeast has become one of the most extensively researched organisms on Earth in its role as a molecular model system and as a crucial constituent in the food industry. It was viewed with equal importance in ancient civilisations although the mechanisms behind its properties were not understood.

Leaven was a portion of dough that contained fermentative microbes, including natural yeasts, and frequently used by ancient communities to make risen bread. To the Hebrews, enslaved by the Egyptians, leaven represented corruptive or egotistical influences. It was omitted from ancient sacrifices and rituals such as the Passover, a ceremony that was performed to ward off evil spirits and the bringers of plague.

The Leaven

Through religious practices, the Hebrews had developed an intriguing knowledge of quarantine and disease control, a quality that the Egyptians seemed to lack. This difference in culture may have contributed to the Exodus, allowing Moses to lead the Hebrews from slavery while the Egyptians were preoccupied with disease. After escaping Egypt, the Hebrews developed the Law of Moses (the Torah) to preserve the unique disease management skills they unwittingly possessed. Two major religious cults, the Pharisees and the Sadducees, evolved from the teachings of the Torah.

As time passed the Hebrews settled in Palestine, but despairingly found themselves again oppressed, this time by the Romans. The Hebrews once more looked towards a redeemer to free them from these shackles, a contender to fulfil this role being Jesus of Nazareth.

The Pharisees, Sadducees and the conquering Romans enjoyed the type of power and wealth that consequently created an underclass of disgruntled people. To these suppressed people, the philosophies of Jesus seemed an attractive alternative to the corrupt religious doctrines of the high priests. Not surprisingly, Jesus lost favour with the Pharisees and Sadducees. Not only was he corrupting their followers but also he disobeyed the Torah by associating with unclean people. The religious high priests conspired to dispose of, what was to them, a dissident deceiver.

The festival of the unleavened bread had been preserved to remind the Hebrews of their flight from Egypt. It was during this festival on the eve of the Passover that the last supper took place. Since then, the festival of the unleavened bread has been used to celebrate the body of Christ in the Eucharist. This was another twist to the original Passover ceremony. In the Eucharist, yeast is omitted from bread to demonstrate the purity of Christ's body. Curiously though, his blood is represented by wine also fermented by yeast. Wine, blood, animals, and unleavened bread were all important components of ancient Hebrew sacrifice.

The purpose of this book is not to condemn or favour views held in religion or science but to bring these doctrines together in order to analyse the way in which societies deal with uncertainty and faith. Uncertainty perhaps arising from a loss of faith, brought about by controversies or lack of knowledge. The first and second chapters investigate the objectives of both science and religion in searching for solutions that will resolve uncertainty. By investigating how the fermentative properties of leaven are interpreted in the

Bible, there is a surprising insight into how ancient societies dealt with uncertainty.

Scientific investigators, such as Copernicus, Galileo and Isaac Newton, demonstrated that through the gift of reality and natural laws, society may be able to share the mathematical structure by which God created the universe. Theologians took up science in order to decipher the complexity of God's creation. For instance, Gregor Mendel, an Augustinian monk who was equally committed to science and religious endeavour, established the laws of inheritance.

Tensions between science and religion began to emerge when evolutionary theory challenged the Christian concept of creation as presented in the Bible. Darwin, also a theologian, produced evidence of a gradual adaptation that had taken place over millions of years. This was in direct conflict with religious perceptions of ready-made organisms that had recently been placed on Earth, only a few thousand years ago. In the molecular era, religion and science are often in dispute over certain issues, especially those that question Biblical concepts, such as evolution, novel genetic techniques, cloning human embryos, and stem cell research.

The third and fourth chapters focus on how science resolved the processes of fermentation through advances in microbiology and biochemistry. It reviews the importance of microscopy in the understanding of disease and discusses the development of knowledge that allowed yeast to be characterised. Theories involving the spontaneous generation of living beings were widely accepted for centuries.

The Greek philosophers Plato and Aristotle in the 4th century BC held views that contradicted a gradual evolution of life forms. Plato argued that there were two worlds; one was real the other imaginary. To them, the variations that were present in plants and animals were merely imperfect illusions of an already perfect form. The discovery of humans at various stages of evolution has diminished the concept that humans were created in their current form or were generated spontaneously. Evolution theories, creationism and spontaneous generation are discussed in these chapters which end by discussing the influence of yeast in the development of western culture.

Chapter five mainly deals with pagan religious practices that were followed in the Old Testament. The Old Testament is derived from four literary sources that span over several decades from 950 to 587 BC. The most authoritative form was thought to be The Pentateuch. The Pentateuch was adopted around 400 BC and consisted of five books from the Old Testament: Genesis, Exodus, Leviticus, Numbers, and Deuteronomy. The Hebrew word for these five books is the Torah, meaning law or teachings. The Pentateuch mainly describes the story of Moses, his birth, teachings through a covenant with God, and ending with his death.

The Passover was an ancient ceremony carried out by Hebrews as a ritual to ward off evil spirits. It was practiced by shepherds, to protect lambs and goats during birth and began at the first full moon of spring. Thereby, explaining why a young sheep or goat was chosen as the sacrificial beast. The Hebrews believed many evil spirits were present at this time and that they would kill newly born animals. The blood of the sacrificed animal was smeared on doorposts to keep away the 'Destroyer'. The Destroyer would most likely be the bringer of disease or plagues. Leaven was not allowed to be eaten or placed anywhere within the dwelling during the Passover.

When the Hebrews left Egypt they continued to treat leaven as a corruptive substance that could displease God. The Book of Leviticus contains a number of rules and regulations that would prevent disease from spreading within the community. Chapter five further discusses diseases that were present in the Biblical era and the procedures used to control them, such as the quarantine of affected individuals and obsessive personal hygiene. It closes by proposing that the Exodus could have been aided by the epidemiological differences between Egyptian and Hebrew cultures.

As with the Old Testament, the New Testament was written during a time of rapid social change. Chapter six discusses the changes in religious attitude that occurred as society gradually abandoned ritual sacrifices and looked for other means to resolve uncertainties.

The Jews were a minority group struggling against the vastness of the Roman Empire. There was confusion and doubt surrounding religious beliefs as the Romans had conquered Egypt and Greece, combining a multitude of different Gods and ideals. Greek philosophy impacted social behaviour influencing education, life-styles, and religious views. Despite these changes, obeying the Torah was still viewed with great importance, in fact, there was a religious court set up specifically to punish those who disobeyed.

The Leaven

The Jews thought that if they did not follow the words of God, as told to Moses in the Torah, they would become slaves once more. The teachings of Jesus of Nazareth did not obey the Torah in the way that the Jews were familiar and, therefore, the religious groups wanted him to be tried in the court. Chapter six describes the social changes that were impacting this era and how they have changed the conception of disease management. It describes the approaches taken by Jesus in dealing with diseased individuals and the subsequent tensions that were created between him and the Pharisees.

In the synoptic gospels of the New Testament, Jesus uses the characteristics of leaven to denote human behaviour or, more specifically, a mutual social conduct that permeates through society, such as, sin or corruption. Symbolically he was warning his followers against false doctrine and hypocritical practices. There were numerous conflicts between the philosophies of Jesus and the Pharisees, particularly in the association with undesirables, sinners, and social outcasts.

According to the Pharisees, Jesus constantly defiled himself by coming into contact with lepers and outcasts and therefore was ritually unclean and in direct contradiction to the Torah. The Torah vindicated sacrificing the few in order to spare the many while the

Greco-Roman influenced philosophies of Jesus demanded that the few should be saved even if this was detrimental to the majority. Chapter six closes by discussing the rituals performed at the last supper that took place during the festival of unleavened bread.

The teachings and philosophies in the New Testament are predominately analogical; they tried to encourage new insights by allowing people to draw comparisons with familiar situations. Parables were used to encourage self-assessment. They were also easy to remember and the stories could permeate through the community. The purpose of chapter seven is to compare the divergent methods by which science and religion are communicated. In particular, how risk and uncertainty are communicated through the media especially in relation to disease management.

The leaven parable can be divided into three component parts: the leaven, the woman, and the meal or flour. Each of these components play a different role in the message being conveyed within the parable and are discussed separately in the text. There are many ways of interpreting this particular parable. Perhaps this is synonymous with the many different ways that the Bible, science, or religion can be interpreted. Chapter seven takes the opportunity to discuss the role of alternative hypotheses in the development of science and religion and the ways in which they are recorded.

Wine is not referred to in the Bible as either leaven or unleavened although it does play a prominent part in sacrifices and rituals. In the Old Testament, it is used in large quantities as part of a daily sacrificial offering that also included animals and unleavened bread. Wine is also offered on the Sabbath and on the first day of the month, where the quantity varies depending on the type of animal used in the sacrifice. Most notably, wine was offered in the daily sacrifice during the festival of the unleavened bread.

Jesus refers to himself as the sacrificial lamb used in the traditional Passover ceremony and the wine was the sacrificial blood. Chapter eight discusses the present debate of using wine as opposed to water or fruit juice in the modern Eucharist. It also discusses how advances in microbiology have changed the social conception of the Bible.

The Eucharist was established to remind how Jesus gave his life in return for his convictions. In many aspects, this modern religious ceremony seems to go against the philosophies of Jesus by its ritual connotations and sectarian exclusion. It perhaps serves more as a means of retaining the ceremonial sacrifices, religious gibberish, and symbolic worship of the Pharisees that were rejected in the teachings of Jesus. The nature of traditions and rituals in relation to science and religion are discussed further in chapter eight.

Not only does yeast serve as one of the most important organisms throughout domestic history, in recent years it has also substantially contributed to biological research. The numerous molecular techniques that have evolved in yeast allow it to make an important contribution to a number of areas in science.

Through studying various types of yeast and other microbes, scientists now know a great deal about the molecular processes involved in cell division, rapid evolution and disease. As a consequence of these discoveries, disease in the 21st century is generally less feared than in the Biblical era, when life expectancy was rarely above thirty-five years.

Chapter nine discusses how advances in microbiology and genetics have helped eradicate many diseases that were endemic in the Biblical era. Hebrews believed that illness and disease were the wrath of God, which is probably why religion played a prominent role in disease management.

Work in yeast genetics has greatly contributed to our understanding of mutation and, therefore, the mechanisms that lead to cancers and some hereditary disease. Yeast is also used as a model system to research ageing. The ninth chapter discusses the emergence of yeast as a molecular tool and the diverse type of research it covers and closes with discussion about the impact of molecular research on society.

In many ways, science and religion persist partly to alleviate the fear of uncertainty, especially where disease management is concerned. To the ancient Hebrews, uncertainty in this area existed possibly through a lack of knowledge and control. It is evident from the rituals described in the Torah that fear preoccupied the lives of the Hebrews and they looked towards religion to search for solutions. Perhaps through practicing rituals they were hoping to control the merciless elements of disease and in some respects this seems to have worked.

Through trial and error, they seem to have created a regime of quarantine and cleanliness that most likely led to disease prevention. They ritually washed utensils and hands, burnt uneaten food, avoided animals with blemishes, isolated people who showed any sign of or had come in contact with disease, and viewed leaven as an impurity that was completely excluded from their living areas during various festivals.

This final chapter compares uncertainty in Biblical times with the uncertainty that is present today when society loses faith in science. There are a number of factors that could be perceived as permeating an influence like leaven: the media, scientists, commercial companies (pharmaceutical, agricultural), academics, and government policy.

Today's society seems to be as fearful of uncertainty as ancient societies were. Uncertainty and fear associated with science were perhaps earned through a series of controversies. The controversy surrounding the Thalidomide disaster in the 1960s is given as an example. A similar uncertainty surrounds the use of the combined MMR vaccine. Public opinion is concerned about these issues because they affect the vulnerable, those individuals that do not have a choice but rely on the judgment of others, who in turn must have faith in organisations such as pharmaceutical companies and governments. This is discussed at length to close the chapter and the book.

"Whereunto shall I liken the kingdom of God? It is like leaven, which a woman took and hid in three measures of meal, till the whole was leavened."

Luke 13:20-21, King James Version

1

Religious and Scientific Truths

Science and religion are often in contradictory hemispheres even though they originate from a similar train of thought. Both are the products of curiosity, they are both concerned with discovery, and both seek answers to similar questions. It is not then surprising that the outcome of scientific and religious investigation, share the same fate, they are both recorded for future reference, in perhaps what is a kind of altruistic obligation to benefit others.

Perhaps one of the most instantly recognised written examples of a documented religious philosophy is the Bible. A colossal documentation of spiritual endeavour, consisting of around seventy books, arranged within two testaments which were written in quite different states of history. The Old Testament emerged during the decline of Egyptian supremacy over the Israelites, while the New Testament was compiled during the height of the Roman Empire.

The doctrines in the Bible were written at a time when countless mysteries underlay biological and natural processes. The nature of several of these processes, such as evolution and disease, were not understood and therefore thought to be the intervention of powerful deities. Not surprisingly, advances in science have challenged the logic behind many ancient ideologies.

Communities in the Biblical era knew nothing about the causative agents of disease. Because of this, the Old Testament is an enticing record of how society coped with the challenges presented by contamination and disease. In this period many extremely unpleasant diseases, such as smallpox and bubonic plague, would have been endemic and have the potential to wipe out entire communities. It is not then unexpected that epidemics were thought to be plagues summoned by supernatural beings, especially as some sectors appeared to have greater vulnerability than others. When Ethiopian soldiers occupied the city of Mecca in 568 AD they were so afflicted by a severe disease that the conflict was forced to end. The migrating population had no acquired immunity to this disease, which was believed to be smallpox.

Smallpox was a viral infection and a major scourge at the time of the Biblical era. The first detailed observations of it were recorded in 910 AD by al-Razi, an Arabic physician. Al-Razi was in charge of a hospital in Baghdad where he wrote a treatise on smallpox. His research gave some accurate epidemiological descriptions of the disease. For instance, he noted that it was seasonal, occurring predominantly in the spring and he also distinguished the symptoms of smallpox from those of measles. The treatise was translated into Greek and Latin enabling Europeans to prepare for the disease as it spread towards their continent and other parts of the world.

It is not clear if smallpox was as virulent a disease in ancient civilisation as it is today, but there were clearly references to dreaded skin diseases and plagues of boils in the Old Testament. Furthermore, facial lesions that could have been caused by smallpox were found on the mummified body of Ramses V, who died as a young adult. As the pustules were still clearly visible, the disease may have been the cause of his death.

Smallpox eventually spread globally as a consequence of migration and trading. The Spanish introduced it to South America in 1507, when members of an enslaved African community spread it to the local population. The disease spread around the Caribbean and in 1520 it was transferred to Mexico. It was believed that a

Spanish crew member was suffering from the disease when one of their ships landed in Mexico. As it was an entirely new disease the indigenous people had no natural immunity or experience in dealing with it. When it arrived from Europe it practically obliterated the indigenous people of Mexico, the Yucatan, and Guatemala.

As with smallpox, bubonic plague also spread globally through migration and trading. The causative agent of bubonic plague is the bacterium *Yersinia pestis*. It is a characterised by dark buboes, fevers, and vomiting. Whereas, smallpox is restricted to humans, bubonic plague could afflict both humans and animals resulting in a higher number of transmissible routes. Recent research believed that the disease originated from Nile rats and was then transferred to black or ship rats via its vector the tropical rat flea.

It is thought that bubonic plague may have become more virulent in the black rat as it seemed to cause more fatalities in Europe than in the countries it originated from, or perhaps this was quite simply because the Europeans had no innate immunity towards the disease; a similar scenario to the one that allowed smallpox to devastate the indigenous population of the Americas. In 439 BC, an epidemic of bubonic plague that originated from Ethiopia arrived in Europe. The Greek, Thucydides, who survived the disease, recorded its progress. It caused the death of one in three people in Athens, including the statesman Pericles, and is thought to have contributed to the fall of classical Greece.

There is no doubt that plagues and skin diseases preoccupied ancient societies. Hebrews believed that illness and disease were the wrath of a powerful deity and therefore religion played a prominent role in disease management. The High Priest had the unenviable responsibility of devising and implementing a programme of disease prevention by compiling and implementing rules and regulations.

The priest shall examine the sore, and if the hairs in it have turned white and the sore appears to be deeper than the surrounding skin, it is a dreaded skin disease, and the priest shall pronounce the person unclean. But if there is a shiny spot on the skin that does not appear to be deeper than the skin round it and the hairs have not turned white, the priest shall isolate the sick person for seven days.
[Lev. 13.3-4]

These regulations were a type of quarantine to prevent the spread of disease ensuring that infected individuals were segregated from the community until they were proven to be clean.

A person who had a dreaded skin disease had to wear torn clothes, leave their hair uncombed, cover the lower part of their face, and call out, 'Unclean, unclean!' They remained unclean as long as they had the disease, and they had to live outside the camp, away from others, as did any other person that came in contact with an unclean person.
[Lev. 13.45-46]

Tuberculosis was another affliction that was very widespread and seemed to occur frequently, evidence of it has been found in human skeletal remains. The bacteria that caused this disease, *Mycobacterium tuberculosis*, are similar to the bacteria that gives rise to leprosy, *Mycobacterium leprae*. Therefore, people who had managed to survive tuberculosis were more likely to have immunity to leprosy, which was thought to be fairly rare at the time despite it being mentioned frequently in the Bible. When leprosy is mentioned in the Bible it probably also refers to other more common infectious conditions such as ringworm, a fungal infection that gives rise to white hairs within infected areas similar to the condition described in the Old Testament.

If anyone has a boil that has healed and if afterwards a white swelling or a reddish white spot appears where the boil was, he should go to the priest. The priest shall examine him, and if the spot seems to be deeper than the surrounding skin and the hairs in it have turned white, he shall pronounce him unclean. It is a dreaded skin disease that started in the boil.

[Lev. 13.18-20]

Another *Bacillus* that caused skin lesions and may well have been the disease referred to as boils in the Old Testament, was *Bacillus anthracis*, the causative agent of anthrax. *B. anthracis* is nearly always fatal when inhaled leading to respiratory failure and septic shock within two to three days. Prognosis is also not good when the *B. anthracis* is digested after eating infected meat. Cutaneous anthrax is the most common form of the disease accounting for 95 percent of all cases, if untreated the fatality rate is 5–20 percent, relatively low when compared to the other forms.

Anthrax legions only occur in exposed areas, such as the hands and face, and are accompanied by local swelling. Ancient remedies for treating the disease included laying figs directly upon the sores. There may have been some benefit in this, as figs have been found to contain a good source of natural antibiotics and vitamin A. Other natural materials with antibacterial properties, such as sesame oil and wine, were also frequently used to dress wounds. The Hebrews probably discovered that these materials had beneficial properties through experience.

The development of microscopic techniques has revealed the causative agents behind several diseases, while radiocarbon dating and DNA profiling have added new dimensions to archaeological and evolutionary theories. Even so, there are still many uncertainties associated with disease and evolution that science has yet to unravel.

Through studying the molecular processes of yeast it becomes apparent that this simple organism has contributed enormously to the development of civilisation. It is also apparent that science is an ultra-organised entity, arranged into a number of categories and subcategories, preoccupied in what is a global effort to discover the ultimate in scientific truth.

Studying the dynamics of living cells with molecular techniques is a journey fuelled by curiosity, but it also requires collaboration between scientists. Each molecular technique is a consequence of scientific investigation that was recorded in order that future researchers may utilise the data produced. Although a truth in science is normally supported by physical evidence scientists still require some elements of faith, as they rely a great deal on evidence provided by others. Once research has been published it is unusual for it to be knowingly replicated by another scientist unless it is necessary to do so in order to discover further truths.

The Biblical text provides valuable information of the historical route that yeast has taken to finally arrive in the hallowed halls of molecular biology. In fact, science and religion seem to have both emerged through the need to address similar uncertainties, they have co-evolved in a search for truths. The purpose of this book is not to condemn or favour views held in religion or science but to bring these doctrines together in order to analyse the way in which society deals with uncertainties. Uncertainties that possibly arise from lack of knowledge or evidence.

By investigating how yeast, as leaven, was interpreted in the Bible there is a surprising insight into how past societies dealt with fear and uncertainty. The biological processes behind leaven were not understood and it was frequently associated with adverse events. Perhaps because of its ability to expand and become contaminated, it was often used to symbolise corruptive influence. There are parallels to how uncertainty was viewed by society in the Biblical era with views towards science and molecular biology today and, in some instances, there may be a legitimate argument for this comparison.

Despite lack of knowledge surrounding biological processes, the Bible still manages to relay the message that by changing behavioural patterns quality of life can be improved. One purpose of science and religion could therefore be to raise this awareness. Occasionally, however, both science and religion can step out of line to become socially threatening. In the Bible, this kind of corrupt behaviour is compared metaphorically to the permeating effect of leaven in dough. At that time, the properties of leaven were not understood and would have been shrouded in mystery.

The Leaven

Through an accumulation of scientific knowledge, it is now known that leaven is a substance that contains microbial organisms, such as yeasts, that ferment carbohydrates in the dough, producing carbon dioxide and alcohol as by-products.

The process of fermentation became realised over time through countless scientific experiments; a relentless thirst for enlightenment that led to further questions and further experiments to answer those questions. Science sought answers not only to what caused the process of fermentation, but also how this was accomplished and for what reason, furthermore, could this process be exploited for some other purpose?

Through tiny progressive steps rather than a giant leap, yeast has become a commercially important domesticated microorganism and a well-established tool in molecular research. The natural process of leavening or yeast fermentation is now well understood and can be manipulated scientifically in laboratories.

Genetic engineering has advanced to such a level that mere mortals can now achieve processes in days that would have taken natural selection decades to accomplish. However, this knowledge is not shared by all and is still feared by communities who cannot influence the outcome of scientific exploitation. It is possibly the fear of corruptive or erroneous influences behind this scientific manipulation that currently cause social concern.

Uncannily, the philosophical message that is conveyed through leaven in the Bible, over two thousand years ago, could well apply to corruptive influences within modern scientific institutions today. Sometimes, especially through the efforts to secure material needs, other motivation besides scientific enlightenment drives research. This can lead to the controversy that reveals flaws in regulatory systems. Ironically, fear of uncertainty is still present, despite the wealth of knowledge that we now have.

In both science and religion, accuracy in recording information requires a certain amount of faith, especially when confronted by uncertainty. Take climate change for instance, none of us, as individuals, can measure with any certainty that the Earth is gradually warming and that this is the consequence of carbon-rich gases. Yet, many of us believe that this is the case through watching or reading media reports, most of which may or may not be based on legitimate scientific findings. As a consequence of these reports, we are also assuming that all change, created by a shift in climate, will result in global catastrophe even if this is not the case. There is a widely accepted view that anthropogenic change will have a negative effect on the balance of natural resources.

We can only assume that climate change is caused through the accumulation of carbon-based emissions because a fairly limited amount of research has drawn up this conclusion. However, consider if climate change was caused by some other factor besides carbon emissions. In reality, carbon emissions are a by-product of energy production, therefore, the heat created through energy production could be what is contributing to climate change. Furthermore, by comparing our present use of heat-emitting energy to that of a few decades ago this seems to be a reasonable assumption. If this were the case, reducing carbon emissions would have little effect. Solar power, wind power and atomic energy would all be useless in reducing climate change, as only a reduction in heat-producing energy would have any impact. So the preferred scenario is that a human activity that is not going to reduce our quality of life, is causing climate change and we have the ability to control this, to some extent, by changing our behaviour.

Several factors could be contributing to climate change, but all of these factors are based on data presented to us by a relatively small group of scientists working in specialised areas. Many aspects of science and religion, therefore, are based on assumptions that require an element of faith. To this end we rely heavily on the investigation and integrity of others to provide answers and solutions.

Scientific investigation usually comes under scrutiny when unexpected biological catastrophes occur. It is pressured into finding solutions even if the initial catastrophe is not entirely science-related. A good example of this is bovine spongiform encephalopathy (BSE), a mysterious neurological disease, with visibly distressing symptoms that infected British dairy cattle in the 1980s. The way it affected entire herds led the media to speculate that it was some kind of horizontally transmitted contagion.

The public demanded answers to several questions. What was causing the BSE epidemic? Could it be transmitted to humans? Was there a cure? The pressure was on scientific investigators to provide answers.

Fortunately, disease diagnosis was fairly rapid as a significant amount of work on transmissible spongiform encephalopathies (TSEs) already existed. This was a consequence of the Nobel-awarded research of Stanley Prusiner, who in 1972 began investigating a human form of TSE, Creutzfeldt-Jakob disease (CJD). Following a decade of research on human forms of the disease, Prusiner discovered that the causative agent of CJD was a single protein, which he called a prion, a name derived from 'proteinaceous infectious particle'. Further research found that it could be transmitted through contact with spinal nerve tissue or blood-related products and that there was no known cure.

The media was quick to criticise how the British Government, and their appointed scientists, had conducted risk reporting and disease control measures. The mishandling of the crisis provoked controversies because it was influenced by a number of conflicting interests. The agricultural community and commercial retailers were suffering financially. While, health officials and consumers were concerned about public safety. The Government had to find a mutual accord that would satisfy all parties. This conflict of interest led to anxieties that attracted media attention and fuelled public debate.

Spurred on by public concerns and armed with available scientific knowledge, the British Government and agricultural community eventually managed to control the BSE epidemic. There was some evidence that TSEs could be associated with the consumption of diseased tissues, but none that showed that the bovine variant of this disease (affecting members of the *Bovidae* family, which includes sheep and cattle) could cross the species barrier to humans. This gap in knowledge was gruesomely filled when a variant of the disease, vCJD, emerged in humans.

The number of people physically affected were fewer than anticipated, but those few victims suffered the most appalling neurological degeneration. Coverage of this by the media greatly impacted social concern. The overall loss of economy to the British beef industry was extreme, with many export restrictions remaining in place twenty years after the initial outbreak.

If science-related knowledge had not been able to contain an epidemic of vCJD could alternative solutions have been found within religion?

Despite a lack of scientific knowledge in the Bible, by following rules and regulations in the Old Testament, it may have been possible to contain some infectious diseases and zoonoses. The Hebrews were only permitted to eat certain animals, such as those that had cloven hooves, stomachs divided into two parts, and that chewed the cud. Animals that did not fulfil this description, such as pigs, camels and hares, were not eaten. Additionally, animals that were permitted to be eaten but had died of natural causes were not to be touched. In the book of Leviticus there are strict regulations in dealing with diseased animals.

If one of the animals that you use as food dies, then anyone who touches the carcase will be unclean until evening. And if anyone eats the dead meat he must wash his clothes, but he will still be unclean until evening; anyone who carries the carcase must wash his clothes, but he will still be unclean until evening.
[Lev. 11.39-40]

Those individuals that had come in contact with dead animals were isolated to ensure that infections did not spread through the community. This provides an unexpected example of how, by following practices stated in the Old Testament, disease transmission could possibly be prevented thereby protecting the welfare of the community. It seems unlikely, in that case, that diseased carcasses would have been fed to healthy stock animals and therefore incidents involving TSEs may not have arisen.

In general, religion and science are viewed with equal reverence within societies, but pathways to their enlightenment take different routes. In science, hypotheses are presented, they are tested and new paradigms emerge. Procedures are recorded, usually in meticulous detail, so that experiments can be replicated by others, in an ongoing cascade of research. Some experiments may denounce a particular hypothesis while other experiments reinforce it, leading to the image of a progressive state of science that continually challenges and evolves.

In many ways, the Bible can also be looked upon as a book of hypotheses, many of which have been challenged by scientific evidence or changes in social practice while others have withstood investigation and social scrutiny. Perhaps a fundamental difference between scientific and religious investigation is that in science, although preceding work is cited, information is usually resourced from the most recently published work. Whereas, in theology, the origin of the research stems back to the initial philosophies recorded in an officially sanctified book and subsequent research is in justifying these ideals. A contrasting approach to advancement exists between these two disciplines.

Advances in genetic research have led to a fairly sound understanding of how hereditary functions at the molecular level. Based on much of this knowledge, humans now possess the ability to create a living being, albeit, in a far less elegant way than nature could perform a similar task.

Cloning, previously a fantasy in science fiction novels became reality when in the 1990s the public were introduced to Dolly the sheep; a clone produced from the genetic material of an adult cell inserted into an empty ovum. The cloning of mice and cows was soon to follow, making the ability to clone humans, with all its controversial implications, a very plausible event. In the Bible, it is clear that only God could create a human being in this way. In a process hauntingly similar to cloning, Eve was created by God from Adam's rib.

Then the Lord God made the man fall into a deep sleep and while he was sleeping he took out one of the man's ribs and closed up the flesh. He formed a woman out of the rib and

brought her to him.
[Gen. 2.21]

Cloning continues to be an explosive issue that creates conflict between religious and scientific communities. Perhaps because science is not currently in a position to provide answers to all the questions that are being asked about it.

Through their pursuit of knowledge and solutions, scientists sometimes have adopted unorthodox methods that have clashed with the moral objectives of religious organisations. The aim of recorded information in the Bible is to influence and manipulate human behaviour and to this end, it is very effective. However, science is seen to conflict with current religious doctrines on many issues, especially surrounding the sanctity of human life, in all its forms.

Take embryonic research, for example, the scientific rationale is that human embryos are the source of stem cells that have the ability to develop into any form of human tissue. This type of research is required to further scientific understanding by creating the cell lines needed to investigate human disease and its treatment, specifically where stem cells could regenerate lost tissue.

Embryonic research could be socially beneficial, especially to those who may have a need for tissue replacement as a consequence of spinal injuries or organ damage. Replacing this tissue with the patient's own would resolve problems associated with rejection, preventing the need for an individual to take immunosuppressant drugs for the rest of their lives.

In the UK, only since 2002 has it become legally permitted to clone human embryos. Surprisingly, according to a survey conducted by the Economic and Social Research Council (ESRC), many people are unaware that human embryonic cloning has been given legal sanction. From those surveyed, only 25 per cent were aware that it was legal to clone human embryonic cells in Britain, whereas, 47 per cent believed it was illegal. The majority of people believed that the Government would not possibly sanction this kind of controversial research.

Government and charity-funded bodies are largely responsible for financing the scientific research that occurs in the UK. Scientific proposals are submitted to these organisations, it is then refereed and scrutinised by a board of experts. If it meets the criteria proposed by the board it becomes funded. Fortunately, research that is unethical or thought to be of a poor standard does not survive this procedure. In principle, controversies in embryonic research are, therefore, unlikely to occur.

The moral objectives to embryonic cloning are in regard to the destruction of embryos as they have the potential to develop into humans. In addition, there may be the temptation to create genetically manipulated foetuses. The Church and Society Council objected to some proposals on ethical grounds. They were in agreement that embryos surplus to *in vitro* fertilisation could be used for stem cell treatment providing this remains within a 14-day limit, but were opposed to the deliberate creation of embryos for research or to create cell lines to treat disease. In Feb 2016, however, came the first approval in the UK for research using genetically altered human embryos within the 14-day limit. The work on surplus embryos, involves a revolutionary gene-editing procedure that will switch on and off specific genes involved in embryo development.

The Council were also against the creation of genetically manipulated embryos, such as parthenogenetic human embryos, human-animal hybrids, chimeric embryos and human embryos that have been made non-viable. Their main arguments were relating to the long-term uncertainty of such experiments and the lack of ethical controls. Nearly all of these procedures, however, are permitted in the embryos of model animals. Ethical human rights are not extended to other animals in the same way.

It seems likely though, that in the future ordinary cells may be manipulated to behave like stem cells and therefore it is possible that future research would involve very little, if any, embryonic cloning. Consequently, preventing the need to address these ethical issues.

When considering the arguments against human cloning on religious grounds there seems to be a consensus that these advances in science would not be considered in the Bible. Curiously however, as previously mentioned, the book of Genesis [2.21] describes a process not dissimilar to human cloning, whereby God creates a female human from a man's rib. In cloning, somatic or stem cells derived from skeletal tissue are used to create another being that would be genetically identical. Perhaps, rather than denounce human cloning, the Bible portrays the process as a crucial element in continuing human-life on Earth. In religion, as in science, interpreting words effectively will eventually lead to the truth.

2

In Search of Truth

Modern science has become technological and ever advancing in an uncontrolled bravado of satisfying human culture, sometimes irrespective of the consequences. Whereas, religion tends to treads warily, fearful of disrupting the balances of nature that are governed by laws that preceded civilised life. Like a badly matched couple, these two different entities seem to have drifted further apart, but at one time they were closely entwined. Theologians were once dedicated to deciphering God's laws.

Among the phenomenal number of hypotheses proposed by scientists through time, some remain concrete for years or decades, even following the most rigorous scrutiny by thousands of researchers. In science, hypotheses that stand the test of time eventually become laws. Perhaps a good example of this can be found in the work of Gregor Mendel, an Augustinian monk who, in the 1860s, deduced the fundamental patterns of inheritance by breeding varieties of peas.

Mendel was equally committed to religious endeavour and scientific investigation in his quest to unravel the complexity of God's creation. During the era in which he carried out his work, theologians believed that living processes followed predetermined patterns that must be governed by the laws of nature. They believed that if these laws were established, then it may be possible to understand why life exists and to what purpose it served. Mendel was foremost in discovering that physically characteristic traits belonging to an individual were inherited from their parents. This, he achieved by statistically recording the colour and shape of peas.

Mendel reached the conclusion that inheritance followed two principle laws: the *law of segregation* and the *law of independent assortment*. The law of segregation observed that specific traits of an organism inherited from either parent could be passed on to offspring at random. The law of independent assortment observed that each trait was inherited autonomously of another.

Despite having no knowledge of molecular biology or the nature of genes, Mendel's observations arrived at conclusions that have withstood scientific scrutiny for centuries. He was unaware of the impact that his findings were to make in scientific research as they were not fully appreciated until after his death. It was others treading in his footsteps that have endorsed his published results.

Advances in science at the molecular level have further reinforced and enhanced Mendel's experimental work. The biology of molecular evolution now attributes the blueprint of characteristic traits in individuals to deoxyribonucleic acid, more familiarly known as DNA. DNA stores the genetic code which when transcribed leads to the generation of tissues that accumulatively construct a new individual with parental characteristics. Seemingly, our own individual genetic plans exist in order for our inherited characteristics to be probed and scrutinised by well-meaning relatives, but in reality, the existence of genetic diversity strengthens the gene pool.

Most of the findings relating to DNA were discovered fairly recently in the last half of the 20th century. The immense speed of this research has undoubtedly been accelerated by the evolution of the mechanically engineered equivalent of the human brain, the computer. Not only can calculations be accomplished in seconds, information and results can also be distributed among the science community at a similar speed. Answers are now virtually received before the questions are even asked.

If Mendel had carried out his experiments in this era, the results would have been published online to be distributed among an international readership within weeks of being written and he would have been fully aware of their impact. Scientists still search for the meaning of life, but now not necessarily for religious reasons.

Before the advent of personal computers, in the 1980s, research moved at a comparatively slower pace. In the 1940s, scientists knew vaguely that chromosomes consisted of protein and DNA, but assumed that proteins transferred hereditary information and that DNA had a structural role merely to provide a framework for the cell and these proteins.

The view that DNA was merely cellular scaffolding was overturned when Avery and co-workers discovered that purified DNA taken from infectious bacterial cells could be transferred into non-infectious cells, in turn, rendering them also infectious. In 1952, Hershey and Chase reinforced this finding by devising an experiment that distinguished the DNA molecule from those of protein. This was achieved by labelling the different molecules with radioactive isotopes and then determining which component was actually transferred into bacterial cells by measuring the radioactivity emitted from them.

Following this initial research, the importance of DNA was becoming realised and the interest in the molecule surged. In 1953, Watson and Crick deduced the structure of DNA from an X-ray diffraction photograph made by Rosalind Franklin.

In a paper published in Nature, Watson and Crick proposed that the DNA molecule was a double-stranded helix and that its replication was semi-conservative with one strand behaving as a template to give rise to an identical daughter strand. These discoveries have been reinforced by further research and have resulted in the complete characterisation of the DNA molecule. The central dogma of molecular biology proposed by Francis Crick was that DNA contained a genetic code, organised into genes, that was deciphered into proteins by ribonucleic acid (RNA).

Many discoveries in molecular biology have been pioneered in model organisms that are easily and economically grown and, most importantly, are unlikely to threaten the life of the investigator. Such model organisms include bacteria; yeasts; rodents;

amphibians; and several plant species, such as *Arabidopsis* (a small flowering plant), members of the potato family, grasses, and, following Mendel's example, legumes, such as. peas or beans. From all of these species, yeast was affectionately dubbed the workhorse of molecular biology, playing a prominent role in the greater understanding of biological science at the molecular level.

Several characteristics of the DNA molecule and genetic inheritance have been researched through first observing the physiology of yeast mutants. The impact that yeast has had in molecular research was largely due to the emergence of apparatus that could visualise the microscopic universe in which it resides.

Early researchers, on first observing these miniscule yeast cells, thought that they just materialised from substances in their immediate surroundings, this led to the dubious theory of 'spontaneous generation'. In many ways they were correct, yeast cells do rely on nutrients from the immediate environment in order to multiply. However, even the tiniest of cells requires a precise organisation that takes centuries to evolve.

A yeast cell is now known to contain over 6,000 genes encoded by DNA that is organised within 16 chromosomes. Not all the DNA provides the genetic code to synthesise components for the cell, some plays a structural role and also adds to overall mass; rather like packaging material it helps to organise the DNA into chromosomes. DNA containing genetic information is neatly condensed into these chromosomes which are stored in a nucleus surrounded by cytoplasm enclosed within a cell membrane protected by a carbohydrate cell wall. Hundreds of proteins and metabolic pathways are involved in maintaining the homoeostasis of a yeast cell.

Even the simplest of cells are derived from a great deal of natural complexity. Today theories of spontaneous generation appear ludicrous, as through microscopy, the precise mechanisms which lead to cell division have now been realised. A long process of evolution must have given rise to this degree of intricacy. Such complexity could not have been generated purely by chance. In the order of the Universe, every process is known to follow certain physical laws, where biological events occur at random rather than by chance. As Stephen Hawking eloquently said:

> *The whole history of science has*
> *been the gradual realisation that*
> *events do not happen in an arbitrary*

> *manner, but that they reflect certain*
> *underlying order, which may or may*
> *not be divinely inspired.*

One major difference between scientific and theological theory is that scientific hypotheses result from physical rather than spiritual observations. They can be challenged by subsequent experimental investigation or re-examination. For instance, many of Mendel's laws and hypotheses, concerning genetic inheritance, have withstood this kind of scrutiny. In contrast, the theories of spontaneous generation presented by Antoine van Leeuwenhoek and several of his contemporaries, in the 16th century, were eventually disproved.

Leeuwenhoek was a draper, chamberlain and wine-gauger who specialised in making high-quality magnifying lenses. He constructed an early form of microscope with a hand-ground lens that, although technologically advanced for that era, could only magnify specimens by about 250 times their natural size.

Anyone who has observed pond water under a microscope during a science class will be aware that it contains a myriad of darting and spinning life forms of every description. These would have appeared astounding to the uninformed mind; it would naturally be assumed that these miniature life forms would eventually grow into something much larger. Leeuwenhoek used his apparatus to observe blood, serum, semen, and other body fluids, and called these miniscule life-forms 'animalcules'. He perceived that animalcules had arisen spontaneously and were, in fact, microscopic extrapolations of larger entities. Most notably he imagined that spermatozoa were tadpole-like cells that contained their own circulatory and nervous systems.

In the latter part of the 16th century a number of scientists, including Leeuwenhoek and Nicolaas Hartsoeker, published drawings of sperm which they believed to be miniature versions of humans a theory known as 'preformation'. In these drawings, miniature human foetuses were folded as they are observed in the uterus, but within the heads of sperm. Although through lack of knowledge and the limitations of their equipment, these researchers were incorrect they attempted to give an intellectual framework to what they observed.

Subsequent curiosity and the art of experimentation led to the abolition of these theories, but the discovery of microbes by Leeuwenhoek has cemented his name in scientific history. Incidentally, Leeuwenhoek, it seemed, was also interested in the properties of fermentation. Among the many microscopic structures he discovered were globular bodies, sometimes oval or spherical shaped, in droplets of fermenting beer. These were the first known microscopic observations of yeast cells.

Increased curiosity and the need to obtain better images of microscopic structures eventually led to the development of the light microscope that could magnify specimens up to 1500 times the original size. Visible light that passes through a specimen refracts as it travels through a lens thereby enlarging the final image. In the 1930s, improved microscopic resolutions were obtained by using a stream of electrons instead of light waves; an electron microscope can magnify 1000 times greater than the light microscope.

Using electron microscopic techniques, it was possible to visualise viruses so tiny that they were able to infect bacterial cells. These viruses were called bacteriophages. Bacteriophages are alien-like particles that can inject their genomes into a host cell. The viral genetic material is replicated by the hosts enzymes to produce hundreds more of the tiny particles. These particles eventually burst out of the cell, killing it in the process. The process by which this happens has been exploited by scientists to create molecular tools that replicate DNA in bacterial cells.

Being able to see images beyond the scope of natural sight has greatly enhanced scientific and medical research. Visualising cell functions have removed the uncertainty that would have obstructed the advancement of many theories and hypotheses. Improvements in microscopic and genetic techniques have revealed that there is far more to the natural world than first imagined. The advent of photography meant that these findings no longer had to remain in the lab or as drawings within books, an accurate visualisation of experiments and specimens could accompany written diagnoses, thereby increasing the validity of findings.

Media technology allowed scientific results to become widely accessible and distributed internationally. Humans can now see beyond their natural ability and realise that billions of organisms exist in the microscopic biosphere. Additionally, the causative agents of many diseases are no longer a mystery. Yet, along with these innovations remained the sinister irony that organisms this tiny could still impose more of a threat to humanity than those with a far greater mass. Humanity has not underestimated this threat and is slowing winning the war against the threat of extinction through disease. In 1970, the World Health Organisation (WHO) announced the complete eradication of the smallpox virus. Societies no longer had to adopt the extreme behavioural changes stipulated in the Old Testament in order to avoid the spread of disease.

Unlike scientific theories that can be endorsed by experiments or visualised by microscopy, religious theories are not physically tangible. They are a form of spiritual experimentation that cannot be vindicated by quantitative evidence but are imposed by the democratic consensus of a responsive sector of society. Once recorded, religious ideas can be interpreted in several ways that can be influenced by current events or social responsibility. There are many different religions, all of which focus on predetermined beliefs and rituals usually assembled for perpetuity in a sacred book.

Contributors to the Bible used many techniques to make philosophical ideas accessible to the general reader. In the New Testament, many teachings take the form of parables, a mental tool used to illustrate complex psychological theory in an accessible format. This enabled philosophy to become universally read in a way that had a profound effect on cultures and people. In fact, the impact of Biblical writings can influence the direction that science takes in society. There is a universal conviction that the Bible was written under the direction of God and that for a passage to be included, it must have some spiritual significance.

When these passages are interpreted in the 21st century, Bible teachings can often appear barbaric. For instance, in the Old Testament, animals and newborns are sacrificed to appease God.

Give me your firstborn sons. Give me the firstborn of your cattle and your sheep. Let the firstborn male stay with its mother for seven days, and on the eighth day offer it to me.
[Ex. 22.29]

It seemed the Old Testament attempted to manage social behaviour by fear and power. In order to ensure that the rules and regulations of the Old Testament were followed, leaders in the form of priests were appointed by Moses through the will of God. The priests had the power to inspect and regulate the Hebrew community, exiling those that imposed a threat and punishing individuals, often by death, who opposed the ceremonial practices; practices that had been sanctified by the highest authority.

Teachings of the Bible have been interpreted in different ways by the numerous religious factions that have evolved from them, such as Jehovah's Witnesses, Methodists, Protestants, and Roman Catholics. Differences in socio-politics are observed because there is a lack of consistency in how religions interpret these teachings. They have their own sets of doctrines requiring certain beliefs, which in some cases result in hostility and disagreement. For instance, in 1945, Jehovah's Witnesses introduced a blood ban. They refused blood transfusions as they believed the Bible prohibits the ingestion of blood. In the Bible, blood was considered to contain the life-force of all living beings.

The life of every living thing is in the blood,
and that is why the Lord has told the people of
Israel that they should not eat any meat with
blood still in it and that anyone who does so
will not be considered one of his people.
[Lev. 17.14]

In many ways, they were correct to assume that blood is necessary to retain life because it transports the gases necessary for cellular respiration around the body. If an animal loses too much blood through injury, it is starved of the oxygen necessary to produce energy and dies.

In addition to oxygen, blood provides the means of transporting other substances around the body including inorganic electrolytes, nutrients, metabolic waste products, hormones, proteins and antibodies. It is, therefore, a bountiful source of life-enriching substances that can attract disease-causing agents, such as blood-borne viruses, parasites and bacteria. These organisms have managed to find various ways to evade the body's hostile immune system to exploit these resources. For instance, trypanosomes, worm-like microbes that cause sleeping sickness, frequently change the molecular composition of their coats to remain undetected by the immune system. It was perhaps therefore not beneficial to eat anything that still contained blood, especially if the animal had died naturally. In fact, in the Old Testament, an individual was labelled unclean if they had eaten an animal that had died through natural means.

One of the most common blood-borne diseases of the Biblical era was schistosomiasis, a condition caused by a blood fluke parasite endemic to Egypt and other Middle Eastern countries. After maturing in snails, the adult form of the blood fluke invades an animal host, it travels through the circulatory system until it reaches the blood vessels of the intestines where it lays eggs that are transferred back into the environment through faeces. The disease would have been transmitted through contaminated waters used to irrigate crops and, therefore, was probably quite prevalent. Today, even with increased awareness, over 200 million people suffer from the effects of this disease. It was equally common in the Biblical era.

Recently discovered ancient papyri dating back to about 3000 BC offer various methods to alleviate the symptoms of schistosomiasis, which were anaemia, digestive problems, and reduced disease resistance. The creators of these papyri were, of course, oblivious to the fact that many diseases could be attributed to the parasitic actions of microbes that were sourcing their energy supply from humans. It is understandable that Hebrews could have thought consuming blood from animals that had died from natural causes could lead to humans acquiring the same characteristics. Consuming blood contaminated by flukes or bacteria would inevitably lead to the animal or person acquiring the same symptoms through illness. Through ritually avoiding blood, they were unwittingly contributing to a form of disease prevention.

By refusing blood transfusions, Jehovah's Witnesses have inadvertently protected themselves from infectious blood-borne diseases. However, the blood ban has led to legal conflict whereby intervention is deemed necessary to protect the interests of a child that may require a life-saving blood transfusion. In some ways, this is also a conflict between science and religion as blood transfusions were developed through advances in science. This is another instance where science and religion are in opposing Universes even though their aims are both to enhance the quality of life.

Through different religions, Biblical philosophies are used to control the thoughts and actions of individuals by a labyrinth of hierarchy, power, and rituals. This has often led to a confusing view of Biblical teachings, as there seem to be different methods to search for truth and there is certainly variation in interpreting what the truth is.

In the Old Testament ritual seems to take on a greater significance and is viewed as an important way to control the elements of nature that appear to be governed by God's will. In contrast, the New Testament seems to be concerned about dispelling the corrupt influence of many religious ceremonial activities that materialised from the teachings of the Old Testament. Differences between the philosophies behind the Old and New Testament lead to two different and opposing interpretations of the term 'leaven'.

Leaven in the Old Testament is only used to symbolise a negative influence such as ego, corruption or impurity and forms an integral part of religious ceremony. Whereas, the term leaven in the New Testament symbolises the dispersal of philosophical thought.

Many of the statements in the New Testament are intended to be figurative or metaphorical and, therefore, could be interpreted in more than one way. Theologians are often accused of treating these metaphors too analytically and reading meanings into certain words or phrases that were originally unintentional. This is perhaps a major problem of presenting philosophical thought as parables or metaphors; although the concept can be easily conveyed, through changes in social history, it can also evolve new meaning. The interpretations of leaven thereby illustrate that the Bible contains two testaments, with quite opposing views and purpose.

As Biblical teachings can influence the way science is interpreted by society, likewise, science can influence the way society responds to the content of the Bible. This is reflected in certain periods of history were science has significantly altered social perception of established ideals. The science researched by Copernicus, Newton and Darwin has had a fundamental impact on the way society views text within the Bible.

The classical Platonic view of scientific research was that all beings had originated from the mind of a Creator and, therefore, rational order must lie in investigating its plan within the natural world. Through this philosophy, 'scientific investigation' and 'belief in a Creator' were inextricably linked.

Copernicus through the rationality of geometry revealed that the Sun, and not the Earth, was the centre of the universe. He proposed that through the gift of reality, society would be able to share the mathematical structure by which God created the world. In the shadow of Copernicus came an expansion of scientific investigation and new innovations, building on the assumption that scientific reality did not impinge on theological certainty. Investigators also discovered that their socio-political status could be improved by uncovering certain knowledge and, therefore, a scientific revolution emerged.

Galileo supported Copernicus's theory that the Earth moved around the Sun although it contravened the view interpreted by the Bible. When his work was published it brought attention to these contradictions, he was subsequently condemned by the defenders of the Catholic Church and forced to recant his support for Copernican philosophy.

Isaac Newton, later in the 17th century, was not met by the same dissension when he revealed the theory behind the movement of objects in time through laws of mechanics. He emphasised that his discovery was made through merely understanding the mathematics of God. By describing the mechanisms of creation through laws, he was testifying to the unimaginable greatness of the Creator. In the light of his discoveries, Newton played homage to the scientists whose research formed the foundation of his work. In a letter to fellow scientist Robert Hooke he wrote:

> *If I have seen further, it is by*
> *standing on the shoulders of Giants.*

By the 18th century, many theologians embraced the richness of the natural world as it gave testimony to the existence of God. Science and religion momentarily joined forces. New scientific hypotheses, such as Mendel's laws, illustrated the predetermined order and structure of the universe. Tensions began to re-emerge when the precise age of Earth's creation and the origin of humanity were disputed by geology and new evolutionary theory. This became evident when Charles Darwin published Origin of Species in 1859. Although his findings questioned the validity of Creation, as described in the Bible, their knowledge and scope were valued by society.

The initiative taken by Darwin encouraged other scientists to follow a different pathway, one that benefited society through knowledge rather than religion. They recognised that this kind of knowledge could be profitable and lead to increased socio-political status. Therefore, the quest to unravel the mysteries of the natural world in the name of religion diminished.

The thirst for scientific truth became and still remains independent of religious enlightenment. Religion is now generally concerned with moral and ethical objectives while science has become a pursuit of knowledge in order to fulfil practical goals or curiosity.

3

Animalcule, mineral, or vegetative

Throughout the centuries, yeast has been an important commodity in domestic and industrial life. Consequently, its fermenting ability has been the focus of many different hypotheses and paradigm shifts. Fermentation was once thought to be the consequence of a chemical reaction by some kind of substance and never imagined to be the metabolic activity of a living organism. During the Biblical era, it would be unlikely that miniscule creatures were conceived to be responsible for fermentation.

Yeasts are abundant in the environment, especially in the soil, where they are transferred, by insects or other means, onto the skins of fruit and animals, including humans. There exists a wide variety of naturally occurring yeasts, from those that cause fungal infections (*Candida* spp.) to industrially important species used in baking and wine production

(*Saccharomyces* spp.). Yeast belongs to the kingdom Fungi and division Ascomycota, derived from the Greek for ascus meaning "sac" or "wineskin", a name inspired by their characteristic spore cases.

In recent times, yeast has become a major player in biological research and is now one of the most studied organisms on Earth. It was the first eukaryote to have a fully sequenced genome, the majority of its genes have been researched and functionally analysed, and many of these genes have analogues in multicellular organisms. It is, therefore, possible to study molecular processes from mammalian systems within a unicellular model eukaryote, making yeast a favourable alternative to animal model systems. Furthermore, the simplicity of its life cycle and the ease in which it is cultured has made it a valuable resource in medical and biotechnological applications. Yeast has made a great impact on 21st century anthropology, which has influenced social behaviour and scientific research. Humanity would probably have progressed differently if this organism ceased to exist.

The term *leaven* originates from the Latin word *levere*, the French equivalent being the noun *levure* meaning to rise. The English term yeast probably originates from the Greek term for fermenting which was *zestos*.

The Leaven

Derivatives from later European influences resulted in the medieval English term *zeest*, which eventually became *yeast*.

The term *leaven*, when mentioned in the Bible, is probably used to describe fermented dough or sourdough; so called because, along with yeast cells, it contained acidifying bacteria that produce lactic and acetic acid, giving the bread a unique tangy flavour. Natural microbial contaminants of milled grains and fruit were the most likely source of alcohol production and leavening in the Biblical era. This microbial flora would have included wild yeasts that were naturally found on cultivated crops. In just one hundred grams of flour there are up to ten billion microbes, and from these about thirty thousand are natural yeasts.

The historical steps taken to elucidate the metabolic processes and characteristics associated with yeast and fermentation follow an intriguing journey of scientific discovery that spanned nearly four centuries. A journey that commenced with the discovery of microscopy in the 17th century and meandered through the complex science of molecular biology in the 20th century. This chapter looks at the history of science in relation to the discovery of yeast, exploring how the Biblical text has influenced the principles and directions of scientific investigation. For instance, the Biblical version of creation greatly differs from scientific theories of evolution. The ensuing debates this creates typically illustrate the divisions that exist between religious and scientific theory.

It is generally accepted that for an enquiry to be viewed as scientific it must involve the gathering of observable, empirical, and measurable evidence. The scientific method involves the collection of this data to formulate and test hypotheses. A number of proven hypotheses, from various published and recorded sources, can then be strung together in a wider context to form theories. The practice of distributing and, therefore, sharing data is often referred to as *full disclosure* as it permits evidence to be scrutinised by others, thereby, allowing the interpretation of results to be challenged.

The Bible states that a divine being must have in some way directed the creation of Life in order to account for its existence on Earth and its complexity. An understandable viewpoint, as even a simple single-celled microscopic organism, such as yeast, is an intricate living structure encoded by over 6000 genes. It is difficult to visualise how DNA originated and how complexity could have happened gradually over time.

Scientific methods in the 21st century would not defend the concept of human origin as presented in the Bible as it is largely based on theories without the support of tangible evidence. Although, the scientific method may possibly have defended the Bible's version of events when it was first written, there being a lack of evidence to refute the statements.

Ultimately, though, it is impossible for anyone living in the 21st century to know, with absolute certainty, how the World was created without the use of a time-machine. There is certainly no compelling argument for the reason there is life on Earth, so various scientific and theological arguments perhaps should be considered before determinate conclusions are reached.

Based on archaeological evidence, it is thought that the Earth was created approximately 4.6 billion years ago and that life originated in the form of bacteria 3.8 billion years ago. Multicellular organisms, in various microscopic forms, are thought to have first existed about a billion years ago and given rise to simple animals, such as sponges and anemones. More complex forms of animals started to appear 550 million years ago (mya). Arthropods are thought to have appeared first, followed by fish, land plants, insects, amphibians, reptiles, mammals, birds and flowering plants, respectively.

The first primates that resemble humans arrived about 2.5 mya with humans close to the present form arriving just 200,000 years ago. Neanderthals disappeared fairly recently, 25,000 years ago; a short time in evolutionary terms, but still a vast time-scale for a human with an average lifespan of 62.5 years to visualise. When the Bible was written the human lifespan was only around 30 years or so.

The Bible condenses the time-scale that life arrived on Earth within the space of a few days. On the first day, God created light which was divided into day and night. The next day, the water was divided from the sky and earth was divided into land and sea. It was not until the third day that life first appeared when God created vegetation. During the fourth day, planets and stars were added with time being divided into seasons, days and years. On the fifth day, God creates birds and sea creatures, commanding them to be fruitful and multiply. Finally, on the sixth day, he added wild beasts, livestock and reptiles, and created humanity in his own "image". The humans were also instructed to be fruitful and multiply, but also to fill the earth and subdue it. In the book of Genesis, the Bible draws a distinction between animals and humans, with emphasis on human superiority.

God said, 'Let us make humans in our own image, in the likeness of ourselves, and let them be masters of the fish of the sea, the birds of heaven, the cattle, all the wild animals and all the creatures that creep along the ground.'
[Gen. 1.26]

There are a number of anomalies in the account of creation within the Bible. We now know that the stars and planets must have existed before the introduction of life. Seasons would be a prerequisite to the existence of vegetation. We are also aware that reptiles existed before mammalian species on account of dinosaur fossil records. Restrictions in knowledge also prevented any reference to the creation of microbes within Genesis, but there are several references to disease and of fermentation processes throughout the Bible.

Ancient civilisations were unaware of what caused many diseases. In general, injuries resulting from accidents or animal bites were understood and treated by various dressings, including sesame oil, wine and balsams, some of which contained naturally occurring antibiotics. In contrast, the mechanisms behind diseases, involving parasites or microbes, were a complete mystery and thought to have been placed in the body by evil forces. Consequently, these illnesses were treated by making the body hostile to the invader through intense cleansing with various noxious substances. Diseases were largely viewed as a punishment brought through sin or disobedience.

The Lord said, 'If you will not obey my commands, you will be punished. If you refuse to obey my laws and commands and break the covenant I have made with you, I will punish you. I will bring disaster on you- incurable diseases and fevers that will make you blind and cause your life to waste away.
[Lev. 26.14-16]

As microbes where viewed as some kind of mysterious, unexplained power they were also used to represent the spread of sin and corruption. For instance, in his first letter to the Corinthians, Paul the Apostle uses the permeating character of leaven to illustrate the spreading of corruption within a community.

You know the saying, "A little leaven makes the whole batch of dough rise." You must remove the old leaven of sin so that you will be entirely pure. Then you will be like a new batch of dough with no leaven, as indeed I know that you actually are. For our Passover Festival is ready, now that Christ, our Passover lamb, has been sacrificed. Let us celebrate our Passover, then, not with bread having the old leaven of sin and wickedness, but with the bread that has no leaven, the bread of purity and truth.

[1 Cor. 5.6-9]

Fermentation procedures and the art of making leavened bread were first introduced by the Egyptians over 4,000 years ago. In ancient Egypt, brewing and baking were practiced together on the same premises. Wooden and plaster models depicting Egyptians working in a brewery and bakery have been recovered from a tomb dated around 1975 BC during the reign of Amenemhat I. Egyptian hieroglyphs that show pictures of dough rising next to bread ovens have also been found.

Until the turn of the 19th century, yeast was supplied in a liquid form very similar to that found at the bottom of beer barrels. Perhaps, this was also how bread was made in early Egyptian civilisation from fermenting beer. Pliny the Elder noted in the first century BC that Gallic and Iberian bread was particularly light because it had been made with froth from the top of beer.

The Egyptians were, therefore, aware that both processes, to prove dough and ferment beer, used the same substance. To the Egyptians, bread was an important commodity used as a type of currency for trading and for the payment of services. In fact, the workers who built the pyramids were believed to have been paid in bread. Bread was also used to pay homage to Osiris the God of grain. The Egyptians believed that Osiris had given them the gift of risen bread to make it lighter to carry, especially in the journey to the afterlife.

In contrast to the Egyptians who revered fermented dough, the Palestinians were extremely suspicious of it. Perhaps, this could be due to the different methods by which the dough was proved. The Egyptians used fermenting beer, whereas, Palestinians would reserve a piece of leavened dough from a previous baking to make the next batch of bread. This dough could potentially be contaminated by harmful microbes which may not have been completely destroyed by the baking processes. When a person died all leaven was thrown out of the house and from all neighbouring houses because Palestinians thought that it may have contributed in some way to the individual's demise, they believed that the angel of death may have thrust his sword into it.

In Britain during the 15th century, as in ancient Egypt, brewing also took place on the premises where bread was baked. The thick layer of dormant yeast cells that sank to the bottom of the brewing or wine vessel was usually referred to as the *lees* while froth from the top of beer was known as *barm*. Barm from a good batch of beer would be reused to make more beer and to ferment bread dough. Barm, the brewer's equivalent of leaven, used to be referred to as Godisgoode because it was thought to be a gift from God. Yeast used in brewing and baking seemed to be received in a different manner to leaven that was used to make sourdough. In fact, during the 17th century, the Paris Faculty of Medicine tried to ban leaven because in the New Testament St. Paul had signified that it was a substance that denoted corruption.

So it seems that fermentation generated through leaven was not viewed in the same way or as a similar process to the fermentation that produced wine. It is clear from passages in the Bible that leaven is used figuratively to symbolise the permeating influence of certain types of human behaviour, especially corruption. Traditionally unleavened bread is used to symbolise the purity of Christ's body whereas wine symbolises the blood of Christ. This is not without some irony, as wine is made in a similar process and perhaps more likely to lead to acts of sin and corruption when consumed than is bread. Indeed, some religions do take this view and at the beginning of the 20th century, alcohol became prohibited for a time in a number of Nordic countries and North America due to Protestant led objections. Alcohol is also prohibited in some countries that follow Islamic laws. The contradictory views relating to fermentation in wine and bread will be discussed further in a later chapter.

Although its meaning still may remain a mystery, life, in itself, is hard work and requires a lot of energy. It is now well established that the initial source of this energy is provided by the Sun in the form of light, which is absorbed by a photosensitive pigment called chlorophyll found in plants and other photosynthetic organisms. The energy is then trapped in molecules of glucose, a carbohydrate compound composed by a series of chemical reactions involving carbon dioxide and water. Plant consumers then transfer the energy stored within the glucose carbon source along the food chain.

When glucose is broken down it produces adenosine triphosphate (ATP), the compound required to release the energy that powers most cellular functions. The most efficient way for an organism to synthesise ATP, thereby releasing energy, is by an oxygen-requiring process called cellular respiration. In humans, oxygen is transferred into the body from the surrounding atmosphere by respiring, it is extracted from air in the lungs by haemoglobin, which is then circulated around the system in the blood. As it contains the oxygen required for anaerobic energy production humans cannot survive without blood. Blood was, therefore, considered of extreme importance in the Biblical era, as it was the substance thought to contain an animal's character and life-force.

Every living thing is a package of consumable energy, but not every organism can boast a sophisticated circulatory system that enables cellular respiration. Microbes and other lower life forms have to adopt fairly basic means to generate their energy. The energy generating processes of yeast cells produce by-products that have been exploited by human civilisations for centuries. One of the ways yeast requires its solar produced energy is by fermentation; a biochemical transformation that converts carbon sources such as glucose or sucrose into energy, producing alcohol and giving off carbon dioxide in the process.

Fermentation is not as efficient in producing ATP as aerobic respiration but enables yeast to convert glucose into energy without the aid of oxygen. Scientifically defined, fermentation is a catabolic process that makes a limited amount of ATP from glucose without an electron chain (supplied by oxygen) producing a characteristic end product, such as, ethyl alcohol or lactic acids.

During fermentation, yeast not only generates energy from the carbon source but it also breaks it down into an industrially and socially important commodity, namely alcohol. Yeast also has the ability to perform aerobic respiration to give off carbon dioxide, but this process does not produce alcohol. Being able to live with or without oxygen is undoubtedly ecologically advantageous to the microbe. Certainly explaining the reason that they inhabited the Earth long before humans did and why they will still be here long after our fragile species has disappeared.

The mysteries surrounding fermentation were once, and to some extent still are, the subject of great scientific endeavour. It was once assumed that the reaction was chemically induced because investigators were unaware that a living thing unseen by the human eye could exist. The yeast commercially responsible for transforming carbohydrate-rich ingredients, like flour and fruit juice, into loaves of bread or alcoholic drinks is predominately *Saccharomyces cerevisiae* also known as baker's, brewer's, or budding yeast. When sugar is plentiful the metabolic route that this type of yeast chooses is fermentation. During fermentation cells multiply rapidly asexually by budding.

Like other simple life forms, yeasts such as *S. cerevisiae* are fully self-contained within one microscopic cell. *S. cerevisiae* cells are round and, providing they are well nourished with carbohydrates, spend most of their life cycle reproducing vegetatively by growing buds. Buds separate from the parental cell when they reach a certain size in order to follow an individual pattern of growth. Upon maturity, these too can start budding, with each cell producing about thirty progeny. The loss of the bud leaves a scar on the parental cell that can be visualised with fluorescent dyes or electron microscopic techniques. The pattern and number of scars can reveal a lot about the condition and age of the yeast cell. Some yeasts do not reproduce by budding, but by forming a cross-wall rather like the mitotic cell division observed in higher eukaryotes. *Schizosaccharomyces pombe* or fission yeast is an example of this. It divides in a similar way to human cells and is therefore used as a model system to study many human diseases, especially cancer.

When all carbon resources are depleted, cells either enter a stationary phase of non-division or produce spores. Budding yeasts can also reproduce sexually. Adjacent cells of opposing mating types fuse together, in response to pheromones, by forming protruded structures called *shmoos*. The end product is a slightly larger round diploid cell that contains two sets of chromosomes; a method of introducing genetic variability into the cell. This diploid cell can either continue budding or enter meiosis to produce four ascospores, known collectively as a tetrad.

Performing meiosis to generate ascospores is a risky business to budding yeast as it has to temporarily stop increasing population size. Therefore, it only faces this challenge when nutrients are low and its survival is threatened. Ascospores and quiescent cells are resistant to stress and can remain dormant for several months, years, decades or even centuries. While dormant they lie at the bottom of the fermenting vessel to form a thick layer of pale brown sediment. Some cells or spores die, but many retain the ability to begin dividing again when conditions improve, for instance when more sugar becomes available. This mode of survival allows them to remain viable in the face of adversity. They are well suited to harsh industrial conditions such as the arid environment that forms the backdrop of the Biblical testaments.

The concept that living organisms produced leaven was not seriously considered until Erxleben, in 1818, proposed that living vegetative organisms could be responsible for the fermentation seen in leaven and barm. Prior to this, in 1680, Leeuwenhoek, with his early microscope, observed yeast cells in fermenting beer. He referred to most of these single-celled creatures as animalcules because they were believed to be immature forms of larger animals.

These first observations of microscopic cells were not further investigated for another century. Leeuwenhoek's contemporaries were largely preoccupied with the argument centred on spontaneous generation, a belief that animals could materialise from other living or mineral things. Before the ground-breaking experiments by Louis Pasteur in the mid-19th century, which showed that excluding particles from sterile broth prevented contamination by microbes, many theorists believed in spontaneous generation.

Pasteur made great advances in microbial research especially when it had industrial connotations. His doctorate thesis was in researching the crystalline structure of two compounds found in fermenting wine, these were tartaric and paratartaric acids. The structures of these two compounds were identical, but in solution, they rotated polarised light in different ways. Pasteur established that this was due to chirality; he discovered that one structure was the exact mirror image of the other, rather like a left and right shoe.

Science has now established that all living organisms only synthesise left-hand amino acids and can only utilise right-hand carbohydrates, left-hand carbohydrate compounds are synthesised artificially. Pasteur suspected that one of the molecules in wine was artificial while the other had been synthesised by a living organism and that it was this organism that was responsible for wine fermentation.

During the time Pasteur was researching chirality, alcohol production was thought to be a chemical reaction. Lavoisier had demonstrated that if a sugar solution was dropped on to heated platinum it produced carbon dioxide, water, and alcohol. It was, therefore, reasonably assumed that the production of wine, beer, and vinegar was simply caused by an unstable chemical chain reaction. By transferring some of this destabilised solution to a vat of sugar and grape juice, the momentum of the chain reaction would continue.

Yeast cells were known to exist in fermenting wine but were just thought to be an incidental by-product. In the 19th century, French wine production was inconsistent because spoilt wine contributed to great economic loss. Pasteur was asked to research a problem concerning lactic acid contamination in beetroot fermentation. Pasteur noticed that whenever fermentation took place yeast cells were present. He also noticed that when lactic acid was produced smaller rod-shaped microbes appeared. In addition, he observed that compounds other than those formed through the degradation of glucose were present and these tended to be asymmetric.

Pasteur deduced that living cells were responsible for wine fermentation and contamination. He also established that if the wine was heated before fermentation commenced, then the microbes were killed and the wine remained free from contamination. The procedure of heating to sterilise came to be known as *pasteurisation* and is today applied to many foodstuffs.

One of the most notable applications of pasteurisation, and perhaps most beneficial as far as health management is concerned, has been in sterilising milk. Milk, as a rich source of protein, was at one time infected by many pathogenic bacteria, including those responsible for common diseases such as tuberculosis and brucellosis. It was through continual development of his knowledge in microbiology and sterilisation methods that allowed Pasteur to disprove the spontaneous generation theory. Using sophisticated equipment, he found that he could physically exclude airborne microbes from a vessel containing boiled meat thereby preventing contamination.

Pasteur's many contributions to science especially microbiology did not finish with pasteurisation, he also went on to develop vaccinations against anthrax and rabies infections. The French government funded the Pasteur Institute to allow him to treat rabies victims. His fame transcended the Atlantic to America, where another three Pasteur Institutes were set up.

In 1818 slightly prior to the spontaneous generation experiments conducted by Pasteur, Erxleben put forward a theory that a biological interaction was responsible for fermentation. In the 1830s, renewed interest in this theory led to a number of experiments by Cagniard de la Tour, Schwann and Kützing. They proposed that fermentation was caused by a biological organism that used sugar as a food, excreting waste substances in the form of alcohol and carbon dioxide. Many chemists of this period disagreed with this theory, they thought that the fermentation process was entirely physical and did not involve any biological activity.

Von Liebig proposed that fermentation was completely mechanical involving a substance that continually processed a chemical transformation causing sugars to degrade into ethanol and carbon dioxide. This argument was resolved by Pasteur's work. In 1876 Pasteur published a book, *Études sur la bière*, in which he proposed that microorganisms obtained energy in anaerobic conditions by fermentation. His theories were supported by experimental evidence later supplemented by the work of Meyerhof. The fundamental metabolic behaviour of yeast is now known as the Pasteur–Meyerhof reaction.

A brief return to von Liebig's chemical theories followed the discovery that the cell-free juice of yeast extracted by a mechanical press could initiate fermentation. This cell-free juice was called *zymase* and is now known to consist of enzymes. In fact, the word *enzyme*, derived from the Greek term for yeast, originates from this discovery. This extract was unstable, but still allowed the chemical and catalytic reaction that turns sugar into alcohol and carbon dioxide. It eventually became generally accepted that this chemical reaction was only sustainable in yeast cells. It was this chemical reaction that provided the yeast cell with the energy of life.

In a way both arguments were partially right, the process of fermentation does involve a chemical reaction, but one that is catalysed by organic molecules created by a living organism.

4

The Complexity of Life

Different theories and speculations concerning the creation of organic things occur in every religion, as most feel that the complexity of the natural world could not have arisen by chance. However, many scientific discoveries began to challenge the creation of Life as depicted in the Bible. Perhaps the most compelling of these was the theory of natural selection presented by Charles Darwin in the mid-19th century. His book entitled the Origin of Species created tensions between the church and science because it questioned a popular and largely accepted image of creation.

Religious devotees perhaps saw science as being not only a threat to their faith but to their social acceptance and respect. Science innovation threatened to ridicule the basis of their fundamental beliefs and values. It is, therefore, understandable that there was a need to retain Biblical teachings in some form.

In the 19th century, the paradigm shift that was rapidly evolving scientific theory was too extreme to evoke an equally rapid change in religious faith. In order to fully commit to a belief requires a great deal of conviction. This conviction can be impenetrable leading believers to imagine that an evil being is responsible for any deviancy from a steadfast commitment. Any element of uncertainty in religious belief seems to lead to the evolution of new religious theories to give meaning to situations that are too difficult to comprehend. In the New Testament, an interesting method is used to quell sceptics and doubting critics. Individuals who questioned the ideals proposed by Jesus were thought to be influenced by the devil.

After spending forty days and nights without food, Jesus was hungry. Then the devil came to him and said, "If you are God's Son, order these stones to turn into bread."
But Jesus answered, "The scripture, says, Man cannot live on bread alone, but needs every word that God speaks."
[Mt. 4.1-11; Mk. 1.12-13; Lk. 4.1-13]

This not only discourages doubt from those with religious faith but also prevents others from persuading them away from their convictions. It comes as no surprise, therefore, that scientific hypotheses that question religious beliefs are subject to contention.

Through research by evolutionary biologists, we have now discovered that humans are more closely related to leaven than early civilisations could have imagined. Many of the human proteins involved in the fundamental functions of the cell, such as DNA replication, are conserved in other organisms, even in yeast and bacteria. In fact, some human DNA can be expressed in bacteria and yeast to produce protein. Proteins that are homologous in different species are known as orthologues. DNA processing proteins, such as polymerases, are often found in this category. For this reason, the mechanisms of mutation in mammalian cells can be studied equally as well in yeast cells.

The big question often asked of evolution is why creatures evolve at different rates? Why did all yeast cells not evolve into complex multicellular organisms like humans or even plants? Why are there still primitive cells like bacteria? As George Carlin, the American comedian and social critic, is often quoted as having said:

If man evolved from monkeys and apes, why do we still have monkeys and apes?

The answer is that organisms are continually evolving, they are constantly finding new ways to preserve or obtain energy, therefore, new species could arise at any time and the reason that species do not evolve together is because physical barriers often keep groups separated. So each group has a slightly different gene pool to begin with, but this becomes more diverse over time, forcing each to take a different evolutionary route. As life is perpetually being created and evolving, the consequence is there must always be lower and higher life forms.

It is the diversity of life, or rather DNA, that encourages adaptations. Through a constant competition to obtain energy, diversity provides the resources on which selection can act. For instance, a group of higher species A are infected by lower species B and some are killed. Some A-species are able to survive because they have a mutation that makes them more resistant to B-species, but it also makes them less resistant to physical stress C. A-species flourishes until it encounters C. As consequence of this cycle of adaptations, species A no longer resembles its original form and becomes subspecies D; but the species A, that have not encountered stress C, still exist.

So evolution depends heavily on a changing environment and the ability of an organism to adapt. This allows the evolutionary route to continue in a perpetual cycle of various adaptations until many species of organisms evolve. Darwin called this process natural selection or the preservation of favoured races in the struggle for life.

If under changing conditions of life organic beings present individual differences in almost every part of their structure, and this cannot be disputed; if there be, owing to their geometrical rate of increase, a severe struggle for life at some age, season, or year, and this certainly cannot be disputed; then, considering the infinite complexity of the relations of all organic beings to each other and to their conditions of life, causing an infinite diversity in structure, constitution, and habits, to be advantageous to them, it would be a most extraordinary fact if no variations had ever occurred useful to each being's own welfare, in the same manner as so many variations

> *have occurred useful to man. But if variations*
> *useful to any organic being ever do occur,*
> *assuredly individuals thus characterised will*
> *have the best chance of being preserved in the*
> *struggle for life; and from the strong principle*
> *of inheritance, these will tend to produce*
> *offspring similarly characterised. This principle*
> *of preservation, or the survival of the fittest, I*
> *have called Natural Selection. It leads to the*
> *improvement of each creature in relation to its*
> *organic or inorganic conditions of life; and*
> *consequently, in most cases, to what must be*
> *regarded as an advance in organisation.*
> *Nevertheless, low and simple forms will*
> *endure if well fitted for their simple conditions*
> *of life.*
> [Darwin, 1859]

Evolutionary selection relies on the chance that an adaptation will occur and that it will improve the fitness of a particular organism. This directly conflicts with early Christian beliefs that organisms do not evolve and remain as they were originally conceived.

The discovery of humans at various stages of evolution has diminished the concept that humans were created in their current form or generated spontaneously. Humans are thought to have evolved from primates that first appeared around 5–8 mya, and share similar characteristics to chimpanzees. In fact, 98% of chimpanzee DNA is homologous to human, but one of the greatest anatomical differences is in brain size. The chimpanzee's brain weighs less than half a kilo while a human brain weighs around three times that much.

Archaeological evidence suggests that three or four hominid species lived in the African continent several million years ago. One of the earliest of these is called *Ardipithecus ramidus*. Modern humans, *Homo sapiens*, are thought to have descended from a number of different hominid species including *Australopithecus anamensis*, *Australopithecus afarensis*, *Homo habilis*, and *Homo erectus*. There are other early hominids that could be more distantly related: *Australopithecus africanu*, *Paranthropus aethiopicus*, *Paranthropus boisei*, and *Paranthropus robustus*. The hominid that began to disperse itself around the globe was the bipedal *H. erectus*. This species is thought to have evolved into *H. sapiens* only 200,000 years ago; a blink of an eye in evolutionary terms.

As previously mentioned, common arguments by creationists against evolution theory are if humans evolved from apes why have apes not evolved and how is it possible that life on Earth is so diverse even in similar habitats? One explanation could be that, even though life looks diverse, all of it is encoded from just four nucleotides.

Nucleotides play an important role in several biological functions, including metabolism, but they are mostly known for being the base units, thymine (t), adenine (a), guanine (g) and cytosine (c), in DNA. When arranged into predetermined DNA sequences, nucleotides can have very similar patterns even in organisms that appear morphologically different. In fact, although humans may seem to look very different from chimpanzees there is only a 1.23% difference in nucleotide divergence.

Mendel was foremost in discovering that characteristic traits were inherited. He was the first to record genetic variation. It is this same variation that enables some members of a species to follow different evolutionary routes. Advances in science at the molecular level have further reinforced Mendel's experimental work. The mechanisms of molecular evolution attribute the blueprint of characteristic traits to nucleotide code which, when inherited by offspring, leads to the generation of tissues that cumulatively construct a new individual with parental characteristics. Genetic diversity strengthens the gene pool, allowing the plasticity needed to evolve. If there were no nucleotide diversity between individuals there may be no life as variation gives organisms the ability to survive changing environments.

Comparing the nucleotide variation in protein-encoding genes from a variety of organisms reveals a distinct evolutionary history. This complex science, known broadly as *molecular phylogenetics*, can involve the comparison of huge amounts of electronic data, sometimes reaching many terabytes in volume. The human genome alone is composed of over three billion nucleotide pairs that contain thousands of repeated regions and single-nucleotide variations or polymorphisms (SNPs). The variation in the SNPs from individual organisms is exploited to study evolution. Phylogenetics allows species to be characterised into orders and families, not only through physical similarity but also through the arrangement of their nucleotides.

The application of molecular archaeology has largely denounced the explanation of creation as proposed by the Babylonian type theories recorded in the Old Testament. Several views held at this time have been dispelled by science. For instance, rain was thought to fall from seas separated from the Earth by a dome structure which the Creator called sky.

Then God commanded, "Let there be a dome to divide the water and keep it in two separate places"- and it was done. So God made a dome, and it separated the water under it from the water above it. He named the dome "sky".

[Gen. 1.6-7]

Birds and fish were created on the fourth day while animal life on Earth was created on the fifth. On the sixth day, the Creator placed humans on Earth to control the animals that existed in the land and sea. The order in which the animals appear seems to be fairly logical, fish and birds being further down the food chain than the higher animals with humans at the very top. There is no mention of bacteria and parasites. In evolutionary terms, they should appear before the birds and fish on day three.

Theories involving the spontaneous generation of living beings were widely accepted for centuries. The Greek philosophers Plato and Aristotle in the fourth century

BC held views that contradicted a gradual evolution of life forms. Plato argued that there were two worlds: one was real the other was imaginary. The variations that were present in plants and animals were merely imperfect illusions of an already perfect form. This philosophy was known as *idealism* or *essentialism* and ruled out evolution as organisms were already in the form that they were destined to become.

Aristotle recognised that organisms could be arranged according to complexity this is often referred to as a scale of nature or *scala naturae*. He believed that there was an organism at each scale; species were fixed and no evolution occurred. This view persisted for 2000 years and was widely adopted by natural theologists who thought that the Creator had designed each species for a specific purpose.

Linnaeus in the 18th century adopted a filing system for all these species. He was a natural theologian who claimed that he had developed the classification system in order to reveal God's plan. This he clarifies with the phrase:

> *Deus creavit, Linnaeus disposuit;*
> *God creates, Linnaeus arranges.*

During the 19th century, yeasts were thought to be part of the plant kingdom in the division of Thallophyta because they lacked true roots, stems, and leaves. They were eventually classified as fungi because they do not contain chlorophyll or rely on photosynthesis to create energy. Instead, they live a parasitic or saprophytic existence, living off the carbon sources supplied by other organisms. Like other species of fungi, yeast can also form spores.

Jean Baptiste Lamarck was one of the first biologists who proposed a theory of evolution in 1809. As curator of the invertebrate collection in the Natural History Museum in Paris, he observed that insects changed gradually over the centuries. He thought that microscopic creatures were at the bottom of evolution and that they generated spontaneously from inanimate material. Lamarck felt that creatures evolved towards greater complexity and that higher organisms were aiming towards perfection to become completely adapted to their environment. He proposed that organisms adapted continually thereby some aspects of their physiology grew stronger while others disappeared and that these lifetime improvements could be passed on to their offspring.

Lamarck's hypotheses that inheritable traits could be acquired during a species lifetime have been largely disproved, but his views were revolutionary; he implied that species evolved and that they were not fixed. Darwin's theory of natural selection was to directly challenge the current viewpoint of fixed design. Natural theology was the accepted way of thinking and doing science, each species being allocated a specific niche by a supernatural being.

At an early age, Darwin was already a keen naturalist and obtained a degree in theology at Christ College Cambridge where he became the protégé of the botanist Professor Henslow. When he was 22 he joined the crew of The Beagle, a survey ship whose mission was to chart the South American coastline. During this voyage, he collected flora and fauna while other members of the crew surveyed the coast. He was particularly interested in the diversity of species that were present in the Galapagos Islands and recovered over a dozen different types of finch.

Darwin began to understand through his own work, and that of others, that the origin of new species arose from a distant ancestor by the gradual accumulation of adaptations. He saw this within the beaks of the finches that he had collected from the Galapagos Islands. Each finch had a specialised beak that was adapted to forage for the type of food found on its island of origin. This was visible evidence that selection through environmental factors could contribute to speciation.

Darwin was reluctant to introduce his theories publicly because, being a theologian, he was aware of the controversy that this would cause. He wrote an essay on the origin of species in 1844 that remained unpublished until 1858 when extracts from it were presented to the Linnaean Society. In the same year another naturalist, Alfred Wallace published a paper on the evolution of new species. This prompted Darwin to complete his book the Origin of Species, which was published the following year.

The Origin of Species presented a strong argument for natural selection through scientific evidence and became adopted by evolutionists as the primary text on the subject. At the time it was written, molecular genetics and the true nature of genetic inheritance were still undiscovered, so Darwin adopted a Lamarckian approach to explain acquired characteristics; where strengths acquired during the course of a lifetime could be passed down to the next generation.

Mendel and Darwin worked in a parallel universe. They were both theologians whose discoveries emerged at the same time and resulted from a similar method of meticulous observations. Mendel's work did not cause an upheaval equivalent to that of Darwin's, as it described the laws of nature and did not directly threaten orthodox beliefs. Attentions were focused mainly on the findings of Darwin and Wallace and, as a consequence, the significance of Mendel's results remained unappreciated until the beginning of the 20th century.

The integration of Mendel's findings with those of Darwin's led to a greater understanding of inheritance and evolution, but digressed from the common belief that Earth evolved only a few thousand years ago and that each species was created within a similar time-scale. Similarly, Pasteur's research eradicated the idea that organisms could spontaneously generate by demonstrating that microbes could not grow under sterile conditions.

The task of baking and brewing in earlier civilisations would have undoubtedly been difficult without the knowledge of sterilisation and pasteurisation. In ancient times, leaven or sourdough would have been left to rise in considerably unsterile conditions in a warm temperature. This environment would have been optimal not only for yeast but for all kinds of microbial growth including those that were pathogenic to humans. It is not surprising that leaven was associated with impurity and corruption. Excessive contamination could have certainly contributed to disease.

The desired characteristics of the yeast strains used in brewing and baking are different although they use the same species, *S. cerevisiae*. Brewing yeast needs to have an agreeable flavour and an ability to flocculate so that the wort can settle quickly to achieve clear beer. In order to achieve favourable characteristics, yeast strains are selected through generations so that a specific yeast strain with the desired qualities can be produced. In Darwinian terms, this would be known as directional selection. This type of selection results in a variety of yeast that has been artificially adapted for a particular industrial use. For instance, pizza dough is made with dry active yeast. Its slow fermentation allows the pizza to be shaped with reduced shrinkage after baking.

Yeast for the baking industry is usually supplied as a compressed block because this has longer shelf life. Just 2.5 grams of compressed yeast in 100 grams of flour divides until it reaches a population size of 25 billion yeast cells.

There is no question that yeast has transformed the structure of modern culture. In the food industry, it provides baked goods, yeast extracts, and alcoholic beverages. Now, in the 21st century, there are about 30 yeast factories in the European Union consuming about a million tons of cane molasses per annum. European yeast production alone generates an annual turnover of around 800 million Euros. There are now several forms of yeast: compressed, crumbled, active, instant, dried and genetically modified.

In scientific research, yeast is a major model organism used mainly in molecular biology to discover information about the mechanisms of cellular processes. In fact, early in the 20th century, RNA was called yeast nucleic acid because it was first discovered in yeast.

Regrettably, very few women have been attributed to any of the early scientific discoveries associated with yeast. In retrospect, they were less likely to encounter Leeuwenhoek's animalcule containing sperm or beer during their daily routines, but the reason could also be associated with the status of women within religion. As a consequence, they are largely excluded from early investigations were scientific endeavour was predominantly undertaken by men. Text within the Bible indicates that women were preferred to have a more subordinate role, as revealed in a letter from Paul the Apostle to Timothy.

Women should learn in silence and all humility. I do not allow them to teach or have authority over men; they must keep quiet. For Adam was created first, and then Eve. And it was not Adam who was deceived; it was the woman who deceived and broke God's law.

[Tim (1) 2.11-15]

In subsequent chapters, the portrayal of women in the progress of religion and science will be further discussed.

Recent developments in faster high-throughput DNA sequencing techniques, whereby hundreds of DNA strands are sequenced simultaneously on microchips, have enabled whole genomes of organisms to be sequenced at a fraction of the time and cost. The human genome sequence was completed in 2003 by the use of older methods developed by Sanger in 1980.

It took more than 3,000 scientists 13 years to complete the first human genome, at a cost of over three billion dollars.

Currently, a human genome can be sequenced within weeks for under fifty thousand dollars. Consequently, nucleotide pattern variations from thousands of genomes in a large number of species are being compared at an accelerated rate, using SNPs and other molecular markers, to further refine theories of evolution. A major focus of these studies in human genomes is the identification of variation that gives rise to genetic diseases such as cancer. If a small variation in the usual DNA of the human genome can lead to a disease that alters the survival of an individual then, understandably, it could give rise to variation that can increase chances of survival too.

Perhaps, in the beginning, God created life through nucleotides; its continued existence and complexity depending on the ability to adapt to environmental elements.

Biblical Theory in a Molecular Era

5

The Sacrificial Lamb

It is difficult to comprehend what drives a community to perform rituals and sacrifices. Contending emotions are perhaps fear, uncertainty, need, respect, and gratitude. Whatever the reasons, rituals are still an important part of modern life, for some more so than others, but sacrifices have become obsolete, a distant reminder of our pagan ancestry.

In the era of the Old Testament, sacrifices were a major preoccupation of Hebrew life. It is evident by reading through the chapters that animal sacrifices were carried out to avoid uncertainties and as a form of thanksgiving. They were often accompanied by rituals that were performed according to specific instructions outlined in the books of the Old Testament.

When anyone offers an animal sacrifice, it may be one of his cattle or one of his sheep or goats. If he is offering one of his cattle as a burnt-offering, he must bring a bull without defects. He must present it at the entrance of the Tent of the Lord's presence so that the Lord will accept him. The man shall put his hand on its head, and it will be accepted as a sacrifice to take away his sins.

[Lev. 1.2-4]

The Old Testament is derived from at least four literary sources that span over several decades from 950 to 587 BC. Unavoidably, some information may have been lost or contorted through subsequent translations but the most authoritative form was thought to be The Pentateuch, a word that derives from the Greek language and meaning five scrolls. The Pentateuch was adopted around 400 BC and consisted of five books: Genesis, Exodus, Leviticus, Numbers, and Deuteronomy. The Hebrew word for these five books is the Torah, meaning law or teachings.

The Pentateuch mainly describes the story of Moses: his birth; teachings through a covenant with God; and ending with his death. It begins with the book of Genesis. This book provides an in-depth history of

Moses' pedigree starting with an account of primeval beginnings to how his ancestors came to live in Egypt.

The next book, Exodus, recounts the most important event in Israel's history, the escape from servitude by its people. The Hebrews were led from Egypt by Moses. While Moses is in exile from Egypt for killing a slave master, he formed a covenant with God. Through using Moses as a mediator, God provided laws and commandments that Hebrews should follow to avoid returning to servitude. Leviticus, the third book, contained the rules and regulations for performing religious ceremonies in order to honour God. It includes comprehensive details of how sacrifices were to be performed.

The following are the regulations for repayment-offerings, which are very holy. The animal for this offering is to be killed on the north side of the altar, where the animals for the burnt-offerings are killed, and its blood is to be thrown against all four sides of the altar. All its fat shall be removed and offered on the altar: the fat tail, the fat covering the internal organs, the kidneys and the fat on them, and the best part of the liver. The priest shall burn

all the fat on the altar as a food-offering to the
Lord. It is a repayment-offering. Any male of
the priestly families may eat it, but it must be
eaten in a holy place, because it is very holy.

[Lev. 7.1-6]

The book of Numbers deals with the story of the Hebrews after they left Mount Sinai. It includes details of two censuses taken by Moses, one taken of those surviving the exodus on departing Mount Sinai, and another census was taken a generation later. The final book, Deuteronomy, is a summary of Moses' achievements as the people prepare to occupy Canaan. The main objective of Deuteronomy seems to be in encouraging the people to give thanks to God. This takes the form of a liturgy delivered by Moses to celebrate future harvests.

After you have occupied the land that the Lord
your God is giving you and have settled there,
each of you must place in a basket the first
part of each crop that you harvest and you
must take it with you to the one place of

worship.
[Deut. 26.1-3]

This type of thanksgiving has been conserved through religious tradition and is still used as a sign of appreciation for a bountiful harvest.

Leaven cannot be regarded as an entirely synonymous term for yeast as it was more likely to be a lump of dough contaminated with actively multiplying microbes from a diversity of species; that would have mainly included yeasts but other microbes would have also been present. Leaven would most certainly also have contained pathogenic contaminants spread by animals and insect vectors. Nevertheless, the fermenting characteristics of leaven and yeast are most likely to be very similar and, from an uninformed perspective, the biological process behind the ability to ferment would be shrouded in mystery.

Leaven is still used to make bread in the 21st century, both commercially and domestically, but now it is normally referred to as sourdough. Investigating how leaven was perceived in the Bible gives a surprising insight into the socio-politics of the period. It allows the investigator to come face to face with the uncertainties and fears that were experienced by societies during that period. In particular, certain social groups lacked control over many aspects of their lives and destiny, some comfort may have been derived from following certain rituals and rites. Through participating in ceremony there was a degree of control and organisation. There must have been an element of hope gleaned in believing that these actions could bring about change, whereas ordinarily, uncertainty and fear of suffering were the predominate factors.

During the Biblical era, the most populated areas were in the valley of the Nile and Fertile Crescent, which is the region between Palestine and the flood plains of the Tigris and the Euphrates. Heat, stagnating water and human waste would have provided exceptional breeding grounds for a host of pathogens. Among the health conditions that were thought to be common in this region were those arising from parasitical or insect-borne infections, such as bilharzia, malaria, and trachoma; and viral or bacterial diseases that could have included bubonic plague, smallpox, measles, tuberculosis, anthrax, and cholera.

Through progress in scientific research, we are fortunately more aware of the contaminating nature of microbes and their ability to cause disease. We now know enough about the nature of microbial infection to prevent life-threatening epidemics. Great advances in disease management occurred in the 18th and 19th century through the discovery of effective vaccines. One of these advances was in 1798 when the physician Edward Jenner developed a very successful vaccine against the highly dangerous contagion, smallpox. Smallpox could be easily diagnosed because of its characteristic pustules that covered the body. These pustules were especially prevalent in the extremities such as on the face, arms and legs. If individuals survived the disease, they were often reminded of their ordeal by the existence of very distinctive pockmarks.

Jenner investigated scientifically a familiar rural myth that those who had suffered from a non-lethal disease called cowpox did not contract smallpox. He discovered that if a person was inoculated with matter derived from cowpox lesions or, more precisely, *Variolae vaccinae* they were subsequently protected against smallpox.

Almost immediately countries began immunising vulnerable populations. The initial inoculations were so effective that countries embarked on their own disease management programmes with Bavaria making vaccination compulsory as early as 1807. In the 20th century, a viral derivative of the cowpox inoculation called vaccinia was used in a mass inoculation program that led to the complete eradication of smallpox. The last known case occurred in Somalia in 1977 and in 1980 the World Health Organisation officially announced that smallpox had been annihilated.

Pasteur was another great protagonist in the never-ending war against the microbe. He also developed a range of vaccines against several life-threatening diseases, especially those that could be horizontally transmitted to humans from animals such as anthrax and rabies. In homage to the preceding science, he named these disease-preventing inoculations *vaccines*. Among other achievements mentioned previously, Pasteur also pioneered antiseptic techniques that were introduced by his contemporary Joseph Lister. So in a short space of time between the 18th and 20th century, disease management became a sophisticated science that largely replaced folklore, witchcraft, herbal remedies, and animal sacrifices.

The story of the Hebrews exodus from Egypt is where the symbolic importance of leaven in the Bible is first introduced. At first, it seems that the Hebrews were welcome in Egypt, enjoying a reasonable lifestyle. Amenenhat III is thought to have been the pharaoh that had made Joseph a vizier and allowed them to settle in the Delta. This was during, what archaeologists described as, the Middle Kingdom, which occurred from 2050 to 1786 BC. Joseph was a Semite, the favoured son of Jacob and descendant of Abraham the prophet. His popularity, believed to be a blessing from God, enabled him to become an important Egyptian governor.

Joseph is noted for having dreamlike premonitions, through which he saved Egypt from famine and brought prosperity to the land. Around this period the Hyksos, renown for instigating the use of horse-drawn chariots, may have invaded vulnerable parts of Egypt. The Hyksos are thought to have built a town called Avaris in the Delta area and would have perhaps looked favourably upon the Hebrews, through empathizing with their predicament.

The Bible implies that as time passes the fate of the Hebrews became uncertain when a new Egyptian king, who did not remember Joseph, became ruler. This possibly occurred during the second intermediate period in Egyptian history, between 1786 and 1567 BC. A political change probably took place about this time when Ahmose I is thought to have expelled the occupying population of Hyksos from the Delta region, destroying Avaris and similar towns constructed by them. The Egyptians needed labour to carry out ambitious projects, so enslaved vulnerable Hebrews and other tribal nomads.

Despite being condemned to slavery, the Hebrew population flourished. The Bible records that the new king, fearing the growing population of Hebrews, tried to crush their spirit with hard labour by forcing them to build the store cities of Pithom and Rameses. In theory, this could have been the rebuilding of towns on the sites of destroyed Hyksos cities. The city known in the Bible as Rameses is thought to have been Pi-Rameses an ancient town that would have covered most of Avaris but is now the site of a modern village called Qantir.

As time passed, despite cruel and harsh treatment, the Hebrew population began to flourish. Fearing this growing population, the King of Egypt ordered all Hebrew male newborns to be killed and demanded they be drowned in the River Nile. The validity of these events is supported by recent excavations that have uncovered a large number of graves found at the ancient site of Avaris. Strikingly, over 65 percent of the burials were of children below the age of two.

Ancient Egyptians believed that their Kings were the sons of gods so names derived from this noun are usually joined to the name of a deity, as in Thut-mose or Ra-meses. It is thought that the Exodus took place sometime during the rule of Rameses II, around 1290 BC. Although there is mounting evidence that points to an earlier date, perhaps during the rule of Dudimose, around 1447 BC. Discrepancies could have arisen because the Bible may be referring to Avaris as the city built by the Hebrews, but used the later name of Rameses. Nevertheless, the Exodus most likely occurred during the New Kingdom, between 1567 and 1085 BC. Although new evidence also suggests that the Hyksos may have invaded after the Exodus when the Israelite population had deserted and there was little resistance.

In the Old Testament, the role of mediator between deities, the pharaohs, and Hebrews was entrusted upon Moses. Moses was born to a Hebrew mother who naturally wanted to protect her son from harm. Out of compassion, she disobeyed the Kings order to drown him at birth. When he was a few months old she placed him in a basket made of bulrushes and left him on the banks of the River Nile in the hope that he would be saved. He was rescued from the Nile by the daughter of the Egyptian King and named Moses. The name, Moses, could be derived from the Egyptian noun Moseh, to beget a child.

The Bible, however, gives an alternative interpretation. The Princess who found the child is thought to have named him Mosheh after the Hebrew word for pulled out, mashah, to commemorate the way in which he was saved from the Nile.

She said to herself, "I pulled him out of the
water, and so I named him Moses."
[Ex. 2.10]

Good fortune continued to shine on Moses; he continued enjoying the privileges of this life and was raised in the Pharaoh's court.

As a consequence of his background, Moses sympathised with the oppressed Hebrews. His devotion combined with a sense of morality led him to fatally attack a slave driver in a fit of anger. Fearing retribution, he fled from Egypt to the Sinai Peninsula. During his exile, he encountered a religious sign or illusion, the burning bush, and heard instructions from the Lord.

Moses saw that the bush was on fire but that
it was not burning up.
"This is strange," he thought. "Why isn't the
bush burning up? I will go closer and see."
When the Lord saw that Moses was coming
closer, he called to him from the middle of the
Bush and said, "Moses! Moses!" he answered,
"yes, here I am."
God said. "Do not come any closer. Take off
your sandals, because you are standing on
holy ground. I am the God of your ancestors,
the God of Abraham, Isaac and Jacob."
So Moses covered his face, because he was
afraid to look at God. Then the Lord said, "I
have seen how cruelly my people are being
treated in Egypt; I've heard them cry out to be
rescued from their slave drivers. I know all
about their sufferings, and so I have come
down to rescue them from the Egyptians to
bring them out of Egypt to a spacious land,
one which is rich and fertile."
[Ex. 3.2-8]

It was a common occurrence within many ancient
religions to enter into personal relationships or

covenants with deities through fire or bright lights. In the book of Genesis, Abraham witnesses the presence of a deity through a burning light within animal carcasses and also makes a covenant with the Lord.

When the sun had set it was dark, smoking fire pot and a flaming torch suddenly appeared and pass between the pieces of the animals. Then and there the Lord made a covenant with Abraham. He said, "I promise to give your descendants all this land from the border of Egypt to the River Euphrates."
[Gen. 15.12]

In fact, Moses believes that the God of Abraham, Isaac and Jacob sent him to guide the Hebrews from oppression. When Moses returned to Egypt, speaking through his brother Aaron, he told the Hebrews that the Lord had instructed him to lead them out of Egypt.

The God of your fathers has appeared to me- the God of Abraham, of Isaac, and of Jacob; and he has said to me: I have visited you and seen all that the Egyptians are doing to you.

The Leaven

And so I have resolved to bring you out of
Egypt where you are oppressed, to a land
where milk and honey flow.
[Ex. 3.16-18]

The Biblical account of the exodus from Egypt is the centre of much scientific speculation especially in respect to a series of plagues that were summoned by the Lord through Moses in order to secure the freedom of the Hebrews [Lev. 7-10]:

- In the first plague, Moses, accompanied by Aaron, turned the waters of all the rivers in Egypt to blood. The fish subsequently died and the water was so foul that it could not be drunk.
- Seven days later Aaron and Moses summoned frogs to swarm the land and the Pharaoh's palaces. In desperation, the Pharaoh agreed to Moses' request to let his people go into the desert to make animal sacrifices to the Lord. The frogs then died, were piled into heaps and began to reek.
- When the frog plague subsided the Pharaoh broke his promise so Moses, through Aaron, summoned a plague of mosquitoes.

- The Pharaoh still refused to relent so Moses and Aaron, at the will of God, summoned further plagues. Swarms of gadflies infested the Pharaoh's palace and the houses of his courtiers and into the land of Egypt, but not to the land of Goshen were the Hebrews lived.
- Then a deadly plague killed Egyptian livestock.
- This was followed by a sixth plague in which Moses took soot from a kiln and threw it in the air, when it landed on the Egyptians it brought out boils that turned into sores.
- Still the Pharaoh was resolute. Moses then summoned plagues of hail, locusts, and darkness.
- Even when subjected to fear and threats the Pharaoh was determined not to lose his workforce, so finally Moses and Aaron threatened to kill the Egyptian's firstborn.

There does seem to be a structured and logical process of ecology behind the sequential appearance of each plague.

Moses, first of all, tries to reason with the King of Egypt to free the Hebrews. He asks if the Hebrews could go into the desert for a few days to pray to their God. This attempt fails and causes the Egyptian King to retaliate by increasing the workload of the slaves. Having failed at his first attempt, Moses tries warning the King through a display of awesome magic powers and through summoning a number of devastating plagues.

In Eastern folklore, there is a strong tradition that plagues could be summoned by magical powers. These type of accounts would perhaps be exaggerated through narration. The events summoned by Moses gradually increase in number through subsequent versions of the Pentateuch, from seven in an early account to ten in the complete account. The seven events in the original account included turning the water of the Nile to blood then summoning scourges of frogs, flies, cattle plague, hail, locusts, and the death of the firstborn. In the version revised by the Holy Priests in Jerusalem plagues of gnats, boils, and darkness are added to the list.

What is interesting and the foundation for great speculation is that the plagues, summoned by God through his envoy Moses, made a clear distinguish between Egyptians and Hebrews.

The houses of the Egyptians will be full of flies, and the ground will be covered with them. But

I will spare the region of Goshen, where my
people live, so that there will be no flies there.
[Ex. 8.23]

It is evident from the chapters of Exodus that there is a great void between the customs and social behaviour of the two populations. For instance, when Moses is pleading with the King to let his people go, he asks if they can travel into the desert to make sacrifices of animals so that they will not offend the Egyptians.

"If we use these animals and offend the
Egyptians by sacrificing them where they can
see us, we will be stoned to death. We must
travel three days into the desert to offer
sacrifices to the Lord our God, just as he
commanded us."
[Ex. 8.26-27]

They also lived in different communities. The majority of the 2.5 million Egyptians lived in the Nile Delta, whereas, the Hebrews occupied quarters in the land of Goshen. This type of segregation could have possibly led to a different epidemiology as far as communicable diseases are concerned.

The signs from God summoned by Moses, rivers of blood followed by plagues of frogs, gnats, flies, boils, and the animal disease, may have been a natural course of events. It is not unreasonable to suppose that polluting a river could cause frogs to evacuate then subsequently die from dehydration or disease, the decaying bodies could then lead to an outbreak of flies which would then transfer disease-causing organisms from the rotting carcases to animals and humans.

Naturally, there are a number of interesting scientific hypotheses that try to rationalize these events. Ancient Egyptian papyri give details of an occasion when the water of the Nile turned red and acidic. Fish died, the waters were undrinkable and burnt the skin. The papyri also mention that pests infected the open wounds and that these pests had a larvae and adult stage, corresponding to the gnats and flies in the Biblical version of the plagues.

Volcanic eruptions occurring around the same time are thought to have deposited sulphates in the Nile. The clouds of darkness and hail could be associated with a violent volcanic eruption. Another theory suggests that the blood-coloured Nile could be due to a bloom of toxic phytoplankton that produced a red tide, forcing frogs onto the riverbank. The frogs eventually died from desiccation leaving a plentiful supply of carrion to attract insects like rove beetles and gadflies. Outbreaks of insect would have occurred leading to plagues of disease that would infect animals and humans. There is also the possibility that the Hebrews polluted the Nile with the blood of sacrificed animals and this led to the subsequent chain of events.

In the final warning that Moses gives the King, a plague would lead to the death of every firstborn in the land of Egypt. Firstborn children seemed to have some kind of symbolic significance and, therefore, perhaps participated in unique rituals or practices that made them more vulnerable to disease than the rest of the population.

Some animal diseases that spread to humans can cause miscarriages, the most notable of these being Brucellosis, which attacks animal tissues with a high erythritol content, a sugar found in mammary glands and the uterus. Brucellosis could have been a secondary infection in cattle that were already in ill health following other diseases such as anthrax or perhaps malnutrition or even infections by parasites. If the Egyptians had contracted this disease from cattle it would have resulted in stillbirths in humans and animals.

Moses claimed that the Lord had told him that at midnight every firstborn in Egypt would die. The following paragraphs from the Old Testament illustrate the symbolism and ceremony surrounding social behaviour practiced uniquely by the Hebrews and not by the Egyptians. These passages, taken from recent translations of the Jerusalem Bible, describe a traditional festival called the Passover that had been modified by Moses to prepare for the Exodus.

The Lord spoke to Moses and Aaron in the land
of Egypt:
"This month is to be the first month of all the
others for you, the first month of your year.
Speak to the whole community of Israel and
say: On the tenth day of this month each man
must choose either a lamb or a young goat for
his household. If his family is too small to eat a
whole animal, he and his next-door neighbour
may share an animal, in proportion to the
number of people and the amount that each
person can eat. You may choose either a sheep
or a goat but it must be a one-year-old male

without any defects. Then, on the evening of
the fourteenth day of the month, the whole
community of Israel will kill the animals.
The people are to take a sprig of hyssop, dip it
into the bowl containing the animals blood
and strike the doorposts and above the doors
of the houses in which the animals are to be
eaten. That night the meat is to be roasted,
and eaten with bitter herbs and unleavened
bread. Do not eat any of it raw or boiled, but
eat it roasted whole, including the head, the
legs, and the internal organs. You must not
leave any of it until the morning; if any is left
over it must be burnt. You are to eat it quickly,
for you are to be dressed for travel, with your
sandals on your feet and your stick in your
hand. It is the Passover Festival to honour me,
the Lord."

"On that night I will go through the land of
Egypt, killing every firstborn male, both human
and animal, and punishing all the gods of
Egypt. I am the Lord. The blood on the
doorposts will be a sign to mark the houses in
which you live. And when I see the blood, I will
pass over you; and you shall escape the
destroying plague when I strike the land of

The Leaven

Egypt. You must celebrate this day as a religious festival to remind you of what I, the Lord, have done. Celebrate it for all time to come."
[Ex. 12.1-20]

The Passover was an ancient ceremony carried out by Hebrews as a ritual to ward off evil spirits. It was practiced by shepherds to protect lambs and goats during birth and began at the first full moon of spring. This explains the logic behind the species chosen as the sacrificial beast, either a young sheep or a goat. At this time, they believed a lot of evil spirits were present that would kill newly born animals. The blood of the sacrificed animal was smeared on doorposts to keep away the 'Destroyer'.

The Destroyer was the bringer of disease or plagues. The meat was eaten during a nocturnal family festival and may have included herbs to enhance the smell and make it pleasing to the deity concerned. This is another way by which the community dealt with uncertainty. They did not understand the epidemiology of disease or the causative agents and, therefore, attributed stillbirths to the retaliation of an angry supernatural being.

It is interesting that at the time diseases and plagues were spreading through the Nile Delta area the Hebrews abstained from eating leaven. It was a normal practice of the Egyptians to allow dough to rise in the sun, this would make it a vulnerable target for disease-carrying insects that would inevitably lead to the spread of communicable diseases. By not eating leavened bread for several days the Hebrews were unwittingly protecting themselves from a potential reservoir of disease-causing organisms. In addition, eating only freshly killed meat in the cooler climate of the evening and then completely burning any leftovers would offer further protection from any contaminating microbes. Perhaps as a consequence of this, and segregation in other social practices, the spread of disease was contained within the Egyptian population, permitting the Hebrews to escape at a time when resistance was weakened.

The festival of unleavened bread traditionally occurred the day after the Passover. It is evident that the Hebrews continued eating unleavened bread when they left Rameses heading for Sukkoth, as they had no time to put leaven back into their dough. In an address by Moses to the Hebrews after the exodus, he stated that the day they left Egypt was to be commemorated by the festival of unleavened bread to remind them of the haste with which they departed [Deut 16:3], having no time to put leaven in their dough.

The Lord said, "For seven days you must not eat any bread made with leaven - eat only unleavened bread. On the first day you are to get rid of all the leaven in your houses, for if anyone during those seven days eats bread made with leaven, he should no longer be considered one of my people. On the first day and again on the seventh day you are to meet for worship. No work is to be done on these days but you may prepare food. Keep this festival, because it was on this day that I brought your tribes out of Egypt. For all time to come you must celebrate this day as a festival. From the evening of the fourteenth day of the first month to the evening of the twenty-first day, you must not eat any bread made with leaven. For seven days no leaven must be found in your houses, for if anyone, native born or foreign, eats bread made with leaven, he shall no longer be considered one of my people. You must eat no leavened bread; wherever you live you must eat unleavened bread."

[Ex. 12.15-20]

There was perhaps a greater significance to the exclusion of leaven from the bread. There was perhaps a social motive other than just a symbolic moral significance. Leaven was associated with impurity, possibly because eating fermented dough could occasionally result in illness or food poisoning as a consequence of toxins produced by pathogenic microbes. An inexplicable illness attributed to the leaven would then have been conceived as the work of an angered deity. This would explain why the Hebrews did not eat leaven before they left Egypt, to eliminate any possible risk of displeasing a deity that may hinder their progress.

The Israelites set out on foot from Rameses for Sukkoth. There were about 600,000 men, not counting women and children. A large number of other people and many sheep, goats and cattle also went with them. They baked unleavened bread from the dough that they had brought out of Egypt, for they had been

driven out of Egypt so suddenly that they did
not have time to get their food ready to
prepared leavened dough.
[Ex. 12.37-39]

According to the book of Leviticus, as leaven was perceived to be impure it was often left out of bread in sacrificial offerings to the Lord. However, leaven was permitted in thanksgiving communion, offered in appreciation of the Lord's many blessings. In this ritual the food was shared among the offerors and leaven was often added, perhaps to symbolise the expanse of the harvest. The Harvest festival was performed to celebrate the first harvests of corn and as a kind of thanksgiving ceremony.

You must bring from your houses to present
with the gesture of offering- two loaves, made
of two-tenths of wheaten flour baked with
leaven, these are first-fruits for the Lord.
[Lev. 23.17-18]

The first sheaf of the harvest was offered to the priest; it would be later burnt as an offering. Ring-shaped cakes of leavened bread were eaten in celebrations along with fresh meat and unfermented cakes [Lev. 7.11-15].

When the Hebrews left Egypt they continued to treat leaven as a substance that could displease God. Other references to leaven in the Old Testament further indicate that there is a ritual exclusion of it from sacrificial offerings. For instance, in 'laws governing offerings and sacrifices' within the book of Leviticus, no grain offerings presented to the Lord were permitted to contain leaven [Lev 2:11], but must contain salt, probably as it represents preservation and prevents corruption from taking place by inhibiting the growth of microbes.

None of the oblations that you offer to the Lord is to be prepared with leaven, for you must never burn leaven or honey as an offering to the Lord. You may offer them up to the Lord as an offering of first-fruits, but they must not go as an appeasing fragrance at the altar. You must salt every oblation that you offer, and you must never fail to put on your oblation the salt of the Covenant with you

God.
[Lev. 2.13]

This regulation also applied to grain when it was given as part of a sin offering [Lev 6:17]. The same conditions applied when Aaron ordained his sons as priests, the consecration offering was bread made without leaven [Ex 29:2]. In the book of Numbers, the ceremony to become a Nazarite involved a complicated ritual of animal sacrifice and head shaving in addition to an offering of unleavened bread.

When a Nazirite completes his vows, he shall perform a ritual. He shall go to the entrance of the Tent and present to the Lord three animals without any defects: one-year-old male lamb for a burnt-offering, a one-year-old ewe lamb for a sin offering and a ram for a fellowship offering. He shall also offer a basket of bread made without leaven.
[Num. 6.13-15]

From reading the Biblical text it becomes apparent that bread is an important part of the diet as it was frequently allied to rituals and ceremonies. Moses is instructed through God to always leave an offering of bread in the presence of the Lord. Precise instructions were given to Moses on how a table for serving the bread offering should be made.

Make a table out of Acacia-wood, 2 units long, 1 unit wide, and 1.5 units high. Cover it with pure gold and put a gold border around it. Make a rim round it and a gold border around the rim. Make four carrying-rings of gold for it and put them at the four corners, were the legs are. The rings to hold the poles for carrying the table article are to be placed near the rim. Make the poles of Acacia-wood and cover them with gold. Make plates, cups, jar and bowls to be used for wine-offerings. All of these are to be made of pure gold. The table is to be placed in front of the Covenant Box, and on the table there is always to be the sacred bread offered to me.
[Ex. 37.10-16]

Perhaps this precision offered some kind of order to the Hebrews in what was an otherwise unpredictable life.

The bread that was displayed on this table was called Showbread or Bread of the Presence. There were very specific and precise instructions on how this bread should be placed on the table.

Take 12 measures of flour and bake 12 loaves of bread. Put the loaves in two rows, six in each row, on the table covered with pure gold, which is in the Lord's presence. Put some frankincense on each row, as a token food-offering to the Lord to take the place of the bread. Every Sabbath, for all time to come, the bread must be placed in the presence of the Lord. This is Israel's duty forever. The bread belongs to Aaron and his descendants, and they shall eat it in a holy place, because this is a very holy part of the food offered to the Lord for the priests.
[Lev. 24.4-9]

The 12 loaves of Showbread are thought to represent the 12 tribes of Israel. The loaves were changed every Sabbath, then eaten by the priests who replaced it. The bread that was used as sacrificial offerings was mainly unleavened because leaven was regarded as a corruptive influence or an impurity. It would not be implausible to suppose that ritualistic ceremonies, like the Passover, inadvertently offered protection from disease thereby giving the impression that a supernatural power had spared the participants from destruction. Alternatively, the rituals could have the opposite effect as many involved the letting of blood from animals. Perhaps then, blood transmitted diseases and parasites, such as blood flukes, were a problem for the Hebrews.

The Hebrews believed that a living being's life force and spirit was found in the blood and, therefore, did not consume it but offered it in sacrifices. The book of Leviticus details the processes and regulations by which these rituals should be carried out. Blood from animal sacrifices was often sprayed around altars or used in symbolic ways. For instance, if a high priest had sinned he must kill a bull, take the bull's blood and carry it into the place of worship. He must then dip his fingers into the blood and sprinkle it seven times in front of the incense-altar, smearing some blood on its four corners and pouring the remaining at the base. Animals had to be free of blemish and were, after the appropriate preparation, burnt on an altar, often overnight. The smell of the food-offering was believed to be pleasing to the Lord.

The Leaven

The authoritative and God-fearing message presented by the Old Testament is in sharp contrast with the philosophies of the New Testament. The Old Testament implicates that if the Torah is disobeyed then serious repercussions will occur. Following the various laws of the Torah must have had beneficial effects in an era before antibiotics and vaccinations revolutionised disease control. These doctrines would have offered some protection to a community from diseases that had no known treatment or vaccine.

Currently, controlling microbial contamination and disease still preoccupies society but is no longer shrouded in mystery. Immunisation is now commonplace so that disease in the 21st century is generally less feared than it would have been in the Biblical era, when life expectancy was rarely above thirty-five years.

The removal and burning of leaven are still carried out before Passover in some religions. During the Jewish celebration, it is traditional to hunt for any leaven (also known as chametz) in the house, the evening before Passover, by candlelight with a wooden spoon and feather to dust away and scoop up crumbs to be burnt the following day. Fortunately, blood is no longer smeared on doorposts.

Biblical Theory in a Molecular Era

6

Obeying the Torah.

As with the Old Testament, the New Testament was written during a time of rapid social change. The Jews were a minority group struggling to find a voice against the vastness of the Roman Empire. There were immense confusion and doubt surrounding religious beliefs with conflicting ideals grappling to become the major influence. The Romans had conquered Egypt and Greece combining a multitude of different Gods and ideals in the process. Greek philosophy had a significant stimulus, impacting social behaviour to influence both lifestyles and religious views.

Prior to the Roman occupation, Alexander the Great, a student of Aristotle (356–323 BC), brought Hellenic teachings to the Middle East. In addition to recording information about the culture and natural environment of the countries he encountered, Alexander wanted to disseminate Greek knowledge and values. At the western edge of the Nile Delta, he founded a city named Alexandria that became a prominent seat of learning. Euclid, Archimedes, and Eratosthenes all researched in the museum that he established there.

When Alexander died the region fell into turmoil, with Palestine caught between the constant bickering of Egypt and Syria. Rome, around this time, was heavily influenced by Greek and Oriental philosophies through trading with the eastern countries of the Mediterranean, subsequently, it was receptive to new ideologies. The fact that other societies had their own Gods may have made Romans sceptical about the Gods that they worshipped and many were perhaps open for change.

Once Rome had control over France and Spain in Western Europe it set out to dominate intimidating cultures from the East. It overturned Syria and Palestine, and under the leadership of Caesar Augustus, gained Egypt from Anthony and Cleopatra. Caesar Augustus was emperor of Rome when they invaded Palestine around 63 BC. He was still emperor when Jesus of Nazareth was born, the Hellenic-inclined Herod the Great was King of Palestine.

Caesar Augustus was a powerful and influential Roman emperor. He had mediated in disputes among Roman leaders following the murder of Julius Caesar in 44 BC and had successfully policed trade routes within and around the Mediterranean. It seems he was held in high regard by the people of the Roman Empire who saw him as a saviour-king, constructing temples in his honour as if he were a deity.

In Palestine, Herod built a huge temple to honour Augustus called Sebaste, the Greek equivalent of his name. To encourage the Jews to follow a Hellenistic way of life he also constructed gymnasia, theatres, and stadia. To pay for these ambitious building projects he collected taxes from the Jews. The Jews resented Herod's efforts to bring Greek influence to the district. As a consequence of this, he was always fearful that his position would be threatened and so appointed secret agents to ensure that none of his subjects would be disloyal. In this respect he went to extremes, having his mother-in-law, two of his sons, and a wife executed because he questioned their loyalty.

Upon his death, around the time of John the Baptist, Herod's three sons, Archelaus, Herod and Philip, under the direction of the Roman Empire, distributed his territory among themselves. Herod Antipas ruled Galilee, Archelaus ruled Judea and Philip governed the remaining regions. Rome also appointed a series of procurators to govern the Jews, the most famous being Pontius Pilate.

The procurators were as unpopular as the other occupiers because they resorted to cruelty in order to control the Jews, who persistently refused to acknowledge Greek religions in favour of their own. The Jews believed that if they did not understand and follow the words of God as told to Moses in the Torah they would become slaves once more. It was in this atmosphere of intense suppression that the Jews hoped for a redeemer to free them once more from the trappings of servitude; Jesus of Nazareth a possible contender for this role.

The teachings of Jesus of Nazareth did not obey the Torah in an orthodox or familiar way and, therefore, some religious groups treated him with contempt. The two main sectarian Jewish groups at this time were the Sadducees and the Pharisees. In the New Testament, Jesus uses the characteristics of leaven to denote human behaviour or more specifically a mutual social conduct that permeates through society. For instance, Jesus tells his disciples to beware of the leaven of the Pharisees and the Sadducees [Mk. 8.14-21; Mt. 16.6]. He seems to be symbolically warning them against false doctrine and hypocritical practices.

When Jesus acquired a reputation for performing miracles the Pharisees disputed his authenticity. They doubted his ability to feed thousands with a few loaves of bread, challenging him to execute a similar act as evidence that God approved of him. He could not replicate the same miracle for them. On the journey away from the Pharisees, his disciples began complaining about their lack of food. Jesus defended his inability to perform the miracle to make more bread.

When the disciples crossed over to the other side of the lake, they forgot to take any bread. Jesus said to them, "take care; be on your guard against the leaven of the Pharisees and Sadducees." They started discussing among themselves. "He says this because we didn't bring any bread." Jesus knew what they were saying, so he asked them, "Why are you discussing among yourselves about not having any bread? How little faith you have! Don't you understand yet? Don't you remember when I broke the five loaves for the five thousand men? How many baskets did you fill? And what about the seven loaves for the four thousand men? How many baskets did you fill? How is it that you don't understand that I was not talking to you about bread?

> *Guard yourselves from the leaven of the*
> *Pharisees and the Sadducees!" Then the*
> *disciples understood that he was not warning*
> *them to guard themselves from the leaven*
> *used in bread but from the teachings of the*
> *Pharisees and the Sadducees.*
> [Mt. 16.5-12; Mk. 8.14-21]

This seems to suggest that in fact it was the quality of the leaven that could increase the bread yield or, metaphorically speaking, spiritual fulfilment. It may well be that some leaven had better fermenting properties than others and this could be used as an analogue for comparing the quality of philosophical thought. Here, Jesus compares the unproductive corrupt leaven or doctrines of the Pharisees with the bountiful yield produced by his own.

Politically and religiously, the Sadducees were the most conservative segment of the Jewish population. They were mainly wealthy and aristocratic families that were anxious to stay peaceful with Rome. They strictly followed the Pentateuch and stressed the importance of the Law of Moses (The Torah) in upholding sacrificial rites and regulations governing the priesthood. The main focus of their worship was a temple in Jerusalem where they practised the rites specified in the Torah, many sacrifices were conducted several times a day.

The temple in Jerusalem had been rebuilt, when the Jews returned from exile around 500 BC, according to specific instructions laid down in the Pentateuch. The temple consisted of a series of courts leading to an innermost court that only the High Priest could enter. The innermost court was where God was thought to dwell.

On a daily basis, many sacrifices were performed in the temple. It was always crowded with priests, Jews and people selling sacrificial animals. Money-changers were there to provide the unique coins that were specified by the Torah at an inflated exchange rate. Ritual and ceremony had reached a high level of intensity, even the priests were selected by the stringent criteria of the Torah. Only those that were direct ascendants of the sons of Aaron could officiate at ceremonies.

The High Priest had considerable social status and enjoyed a high amount of authority, being incorporated into government decision-making. He was also head of the Sanhedrin, a court that handled cases that defied the Torah and was recognised by the Romans. Their authority was often questioned by the Pharisees and eventually diminished when the temple of Jerusalem was destroyed in 70 AD.

The disappearance of the temple marked the disappearance of the Sadducees. They were so entrenched in the Temple cult that they could not survive without it. The Pharisees also followed the Torah, but did not concentrate exclusively on the written word in the Pentateuch but also with other writings and books that were being incorporated into the Old Testament. The Pharisees led the Jewish community to recovery following the destruction of Jerusalem and the Temple around 70 AD during the war with Rome.

The Torah excludes certain people from society that are viewed as unclean, such as lepers. The fact that Jesus associated with people considered socially and religiously unacceptable, angered the Pharisees. His radical interpretation of the Torah also alienated him from the Sadducees. The fundamental difference between the philosophies of the Jewish sects and those of Jesus were in obeying the Torah.

The Pharisees believed that righteousness consisted within the Torah, whereas, Jesus believed that the Torah was in itself under the judgment of God. Despite this, it is very clear from his teachings that Jesus did not completely reject the Torah as he was very familiar with its content. For instance, when a follower asked him what he must do to inherit eternal life Jesus referred him to the Ten Commandments.

"Good Teacher, what must I do to receive eternal life?" "Why do you call me good?" Jesus asked him. "No one is good except God alone. You know the commandments: Do not commit murder; do not commit adultery; do not steal; do not accuse anyone falsely; do not cheat; respect your father and mother."
[Mk. 10.17-19]

He also told the man that if he wanted to enter the Kingdom of Heaven he must sell all of his belongings, give the money to the poor and then he would find riches in heaven. When the man walked away dismayed, Jesus turned to his followers and told them that it was easier for a camel to pass through the needle's eye than it was for a rich man to enter the Kingdom of Heaven [Mk. 10.20-25].

There were numerous conflicts between Jesus and the Pharisees, particularly in his association with undesirables, sinners, and social outcasts. In response to the Pharisees criticism, Jesus explained that if you are not sick than you do not need a physician; I came not to call the righteous, but sinners [Mk. 2.17].

According to the Pharisees, Jesus constantly defiled himself by coming into contact with lepers and outcasts and, therefore, was ritually unclean and in direct contradiction to the Torah. The Pharisees believed that eating food without first washing the hands was ritually unclean. They and all Jewish people believed that the Torah instructed them to clean all utensils and food before eating. In response to this criticism, Jesus argued that it is what comes out of a person that makes them unclean not what goes in.

"There is nothing that goes into a person from the outside that makes him ritually unclean. Rather, it is what comes out of a person that makes him unclean."
[Mt. 15.10-20; Mk. 7.14-23]

In many ways, Jesus seems to place the sinners above the self-righteous Pharisees. He explains that those that do not pass judgment on others will be looked upon more favourably in the eyes of God, as described in the Parable of the Pharisee and the Tax Collector.

Jesus also told this parable to people who were sure of their own goodness and despised everybody else. "Once there were two men who went up to the temple to pray: one was Pharisee, the other tax collector. The Pharisee stood apart by himself and prayed, 'I thank you God I'm not greedy, dishonest, or an adulterer, like everybody else. I thank you that I'm not like that tax collector over there. I fast two days a week and give you a tenth of all my income.' But the tax collector stood at a distance and would not even raise his face to heaven, but beat his breast and said, 'God, have pity on me, a sinner!' I tell you" said Jesus, "the tax collector and not the Pharisee, was in the right with God when he went home. For everyone who makes himself great will be humbled, and everyone who humbles himself will be made great."
[Lk. 18.9-14]

More specifically, what Jesus exactly feels about the Torah is explained in the gospel according to Matthew [5.17-20]. He states that he does not want to do away with the Law of Moses and the teachings of the prophets and explains that if anyone disobeys the commandments, and teaches others to do the same, they will be the least in the Kingdom of Heaven. He that enters the Kingdom of Heaven will only do so if they are more faithful than the teachers of the Law and the Pharisees in doing what God requires.

Jesus disagrees with the methods that the Pharisees use to follow the Law of Moses. The Pharisees had evolved a self-righteous approach to obeying the Torah, whereas, Jesus was more concerned with spiritual fulfilment and consciousness. This perhaps explains why he compares the corrupting characteristics of leaven to the philosophies of the Pharisees and the Sadducees. He is suggesting that by obeying the Torah according to their teachings, through rules and regulations, you will lose sight of the message that God was initially trying to convey.

The philosophies of Jesus predominately passed by unnoticed until the last year of his life. His teachings in Galilee had not reached the major religious centre of Jerusalem but when he did arrive in the city he caused a major disturbance. Firstly, he arrived not in a messianic role but in a humble manner riding on an ass, as prophesied by Zechariah. In Eastern tradition horses are associated with war whereas the donkey is a symbol of peace.

I have seen how my people have suffered. Shout for joy you people of Jerusalem! Look, your King is coming to you! He comes not triumphant and victorious but humble and riding on a donkey.
[Zec. 9.9]

Secondly, he forced the vendors and money-changers from the temple in direct conflict with the behaviour of those obeying the Torah.

Jesus went into the temple and drove out all those who were buying and selling there. He overturned the tables of the money-changers and the stools of those who sell pigeons, and said to them, "It is written in the Scriptures that God said, my temple will be called the house of prayer, but you are making it a

hideout for thieves!"
[Mt. 21.2-13]

In the days leading up to the Passover, everyone visiting the Temple to worship or make a sacrifice had to pay a temple tax, apart from the Pharisees, High Priests, and Rabbis of the temple. A special currency was used for the tax which could only be obtained from money-changers, who usually offered unfair rates and charged a fee for their services. Money-changers were often relatives or associates of the Pharisees and High Priests. As a consequence of his disruptive actions, religious leaders and Roman authorities considered Jesus to be a rebel who had the potential to influence a Jewish uprising.

Shortly after Jesus leaves Jerusalem, he sensed that his predicament was precarious and arranged a meal with his disciples. Here, unleavened bread is once again used to symbolise doctrines and philosophical thoughts. The last supper was thought to occur during the festival of unleavened bread, kept to commemorate the Israelites flight from Egypt featured in the Old Testament. It seemed to be Jesus's intention to share the Passover meal with his disciples but there is strong belief that he was in fact executed before the Passover ceremonies were due to take place on the Sabbath [Mk. 14.12-21; Lk. 22.7-13, 21-23; Jn. 13.21-30].

Even though the last supper may not have been a Passover meal it was portrayed as one by Jesus who saw himself as the sacrificial lamb, the wine was symbolically the sacrificial blood, and the unleavened bread represented his body.

While they were eating, Jesus took a piece of bread, gave a prayer of thanks, broke it and gave it to his disciples. "Take and eat it," he said; "this is my body." Then he took a cup, gave thanks to God, and gave it to them. "Drink it, all of you," he said; "this is my blood, which seals God's covenant, my blood poured out for many for the forgiveness of sins. I tell you, I will never again drink this wine until the day I drink the new wine with you in my Father's Kingdom."
[Mk. 14.22-25]

Following the last supper, Jesus and the disciples go to the Mount of Olives where he predicted that his followers will deny they knew him. He recites a passage from the book of Zechariah [13.7-9] in the Old Testament which predicts that God will kill the shepherd and that the sheep of the flock will be scattered.

One of the disciplines, Judas Iscariot, had been promised 30 pieces of silver if he betrayed Jesus to the high priests. Judas led armed men to Jesus where he was speaking with his followers in the gardens of Gethsemane. He identified Jesus to the men by kissing him. Jesus was arrested and brought before the high priests. Following a brief court appearance, he was accused of blasphemy and sentenced to death. When Judas learnt that his actions had condemned Jesus, he repented and returned the 30 pieces of silver, stating that he had betrayed an innocent man to death. In dismay, he took his own life.

Jesus was brought before Pontius Pilate who was reluctant to condemn him because he did not understand the charge, but accepting that he was politically dangerous ordered his execution. Following the crucifixion and resurrection of Jesus, several of his followers began to spread his philosophy. Politically there was a thirst for change. Instead of halting the spread of Christian philosophies, the death of Jesus served to intensify the movement.

Jewish Christians spread throughout Palestine and beyond, establishing themselves in Syria. Early missionaries extended the philosophies to Rome. Where the founding of the Catholic Church was attributed to Simon Peter. Armenia became the first Christian state through the work of Thaddeus.

A major contribution to the spread and writing of the New Testament philosophies was the conversion of Saul of Tarsus, a Pharisee, to Christianity. Previously he had been responsible for imprisoning the followers of Christianity. On travelling from Jerusalem to Damascus, with some prisoners, the resurrected Jesus appeared to Saul in a great light. So bright was the light that he remained blind for three days until his sight was restored by Ananias of Damascus. The apparition and blindness, which may have been a consequence of heat exhaustion, served to show Saul the error of his ways. He changed his name to Paul, the Roman equivalent of Saul, and began to preach in lands around the Mediterranean, especially in Greece where the name Christ from the word Christos, Greek for Messiah, was first used.

Paul established a church in Corinth and was attributed with fourteen epistles in the New Testament. In the following passage, he speaks about immorality within the congregation and again uses leaven as a synonym to describe corruption.

Now, it is actually being said that there is sexual immorality among you so terrible that not even the heathen would be guilty of it. I am told that a man is sleeping with his stepmother! How, then, can you be proud? On the contrary, you should be filled with sadness, and the man who has done such a thing should be expelled from your fellowship. And even though I am far away from you in body, still I am there with you in spirit; and as though I were there with you, I have in the name of our Lord Jesus already passed judgment on the man who has done this terrible thing. As you meet together and I meet with you in my spirit, by the power of our Lord Jesus present with us, you are to hand this man over to Satan for his body to be destroyed, so that his spirit may be saved in the Day of the Lord. It is not right for you to be proud! You know the saying, "A little bit of leaven makes the whole batch of dough rise." You must remove the old leaven of sin so that

you will be entirely pure. Then you will be like
a new batch of dough without any leaven, as
indeed I know you actually are. For our
Passover Festival is ready, now that Christ our
Passover lamb, has been sacrificed. Let us
celebrate our Passover, then, not with bread
having the old leaven of sin and wickedness,
but with the bread that has no leaven, the
bread of purity and truth.
[1 Cor. 5-13]

Interestingly, in this passage, the old leaven is portrayed as sin and wickedness. Perhaps another benefit of throwing out all leaven during the Passover was to ensure that a fresh, uncontaminated batch would be started. Leaven is also used by Paul to illustrate corruption when trying to persuade the Galatians that they only required faith to be right with God and there was no need to rigidly obey the Torah.

"You were doing so well! Who made you stop
obeying the truth? How did he persuade you?
It was not done by God who calls you. It takes
only a little leaven to make the whole batch of

> *dough rise, as they say. But I still feel confident*
> *about you. Our life is union with the Lord*
> *makes me confident that you will not take a*
> *different view and that the man who is*
> *upsetting you, whoever he is, will be punished*
> *by god."*
>
> [Gal. 5.7-9]

In the above passage, Paul was mainly attacking the practice of circumcision. Paul argued that circumcision no longer meant the physical, but a spiritual practice and labelled those that advocated it as false brothers.

The strict ritual and practices of the Torah that were designed to protect the Hebrews and prevent them from returning to servitude were ironically used by the Pharisees and High Priests to persecute Jesus of Nazareth. This indeed led to a leaven effect, whereby, the teachings of Jesus were spread by his followers on his death. Eventually, God's word as delivered by Moses in the Torah became part of the Old Testament and placed alongside God's word delivered by Jesus in the New Testament.

7

The Leaven Parable

Then the disciples came to Jesus and asked him, "Why do you use parables when you talk to the people?" Jesus answered, "The knowledge about the secrets of the Kingdom of heaven have been given to you, but not to them. For the person who has something will be given more, so that he will have more than enough; but the person who has nothing will have taken away from him even the little he has. The reason I use parables in talking to them is that they look, but do not see, and they listen, but do not hear or understand. So the prophecy of Isaiah applies to them. "

"This people will listen and listen, but not understand; they will look and look, but not see, because their minds are dull, and they have stopped up their ears and have closed their eyes. Otherwise, their eyes would see,

their ears would hear, their minds would understand, and they would turn to me, says God, and I would heal them."

"As for you, how fortunate you are! Your eyes see and your ears hear. I assure you that many prophets and many of God's people wanted very much to see what you see, but they could not, and to hear what you hear, but they did not."

[Mk. 4.10-12; Lk. 8.9-10; Mt. 13.10-16]

The teachings and philosophies of Jesus are predominately analogical; he tried to encourage new insights by allowing people to draw comparisons with familiar situations. Perhaps this is why simple foodstuffs and domestic chores feature so many times in the Bible. It is fairly evident that his preferred audience are not the wealthy or powerful so many of the terms he uses are familiar to all levels of social class. By using parables, he is encouraging freedom of thought in an imaginative style that would appeal to this audience. Parables encourage self-assessment, are memorable and others could pass the stories through the community. Moreover, if memorable, they would be more likely to permeate through society.

The parables also tend to be grouped together by subject manner. For instance, the parable of the lost sheep is grouped with the parable of the lost coin and the parable of the lost son [Lk.15]. These parables all deal with the issue of finding belief when it has been lost. In the parable of the lost sheep, Jesus describes how joyous a shepherd is when one of his sheep strays and is found, even though he has several that are not lost. Similarly, in the parable of the lost coin, a woman loses one of her few coins searches frantically for it and rejoices when it is rediscovered. In the final parable a father celebrates when a sinful son repents, he was lost and then was found. Some of the parables are told within the context of life at that particular time in history, perhaps referring to a current or political situation. Therefore, the parables are sometimes difficult to interpret accurately.

The leaven parable, mentioned by Mathew and Luke, is grouped with two parables describing the Kingdom of Heaven, the parable of weeds [Mt. 13.24-47] and the parable of the mustard seed [Mk. 4.30-32, Mt. 13.31-32; Lk. 13.18-19]. In each of these parables, the coming of heaven could be interpreted as the end product resulting from the growth of all these items; the leavened loaf of bread, the weed-free harvest, and the fully-grown tree from a mustard seed.

The parable of weeds describes how the Kingdom of Heaven is compared to a field sown only with good seed. Any weeds that grow represent corruption and are separated from the crop, to be destroyed by the harvesters. In the context of the era, leaven may have had a different connotation. The parable of the leaven consists of just two sentences.

"The Kingdom of heaven is like this. A woman takes some leaven and hid it in three measures of meal until the whole batch of dough rises."
[Mt. 13.33; Lk. 13.20]

To many, this parable is thought to illustrate how the Gospels will slowly permeate through society until all are converted to Christianity. This is in contradiction to how leaven is usually interpreted in the Bible, to symbolically represent corruption.

The leaven parable can be divided into three component parts: the leaven, the woman, and the meal or flour. Each of these component parts plays a different role in the message being conveyed within the parable and are discussed separately in the following paragraphs.

The Leaven

Leaven when used figuratively in the bible is most often used to denote something that is corrupt. The initial conception of it in the parable is of a favourable component in the dough, this implies that Jesus is contradicting its symbolic use in the Bible.

This parable could be interpreted in many ways. The leaven is taken by the woman and hidden in the meal or flour. The leaven may represent the subtle way that evil can permeate through the dough. In this manner leaven still represents something that corrupts, disintegrates and breaks up. The woman is impregnating the pure symbol of heaven symbolised by the meal with evil symbolised by the leaven.

The first indication that leaven was associated with corruption in the Bible was when Lot baked unleavened bread for the angels. Lot did not give leavened bread to the angels because he did not want to offer them anything containing impurities.

But he insisted so strongly that they did go with him and entered his house. He prepared a meal for them, baking bread without leaven, and they ate.
[Gen. 19.3]

The next reference to leaven is in the book of Exodus was when the Hebrews left Egypt. They received a command from the Lord, through Moses, not to eat leavened bread for seven days. Similar sentiments occur in later sections of the Bible when the Lord demands that leaven should not be offered in sacrifices to him.

Do not offer bread made with leaven when you sacrifice an animal to me. Do not keep until morning any part of an animal killed at the Passover festival.

[Ex. 34.25]

Leaven was excluded from any sacrifice because it was thought of as a contaminant that did not reflect sinlessness. Leaven during the Biblical era would have contained many undesirable elements in addition to fermenting microbes. It's unlikely that the primitive baking processes used at that time would have destroyed all pathogenic microbes and, therefore, there was the potential for leaven in bread to transmit diseases. This is perhaps why it earned such a bad reputation and why people tended not to offer it to guests during festivals or at other times.

Interestingly, animal sacrifices were disposed of before they started to decompose or they could become contaminated also. There was the additional fear that the Egyptians would look unfavourably upon the practice of animal sacrifice and there would be consequences for the Hebrews if they were caught with animal remains.

If we use these animals and offend the Egyptians by sacrificing them where they can see us, we will be stoned to death. We must travel three days into the desert to offer sacrifices to the Lord our God, just as he commanded us.
[Ex. 8.26-27]

The words leaven and unleavened occur over sixty times in the Old Testament and nearly twenty times in the New Testament. In every instance, except for in the leaven parable, it is used to denote something corrupt or sinful. Mostly, Jesus uses leaven figuratively in the same way as it is used in the Old Testament to denote corruption. For instance, Jesus compares the doctrines of the Pharisees with leaven.

> *Be on guard against the leaven of the*
> *Pharisees, I mean their hypocrisy.*
> [Lk. 12.1]

In Mark's gospel, the leaven 'of Herod' is added to the above quote [Mk 8.14-15]. According to some interpreters, the leaven of the Pharisees can be interpreted to represent the hypocritical formality and ritual of their beliefs. The leaven of the Sadducees was rationalism and the denial of supernatural events. The leaven of Herod was the consequence of these two doctrines, a departure from God and his teachings to a devotion of secularism and indulgence.

Throughout history, the nature of leaven has led it to be seen with similar connotations. A rabbi reportedly said, "Trust not a proselyte till twenty-four generations, for he holds his leaven." Here leaven is used as a symbol of hostile infidelity. It was also used by the Talmund to signify "Evil affections and the naughtiness of the heart." The ancient interpretation of leaven by the Greek historian, Plutarch, presented a figurative meaning that had similar connotations:

The Leaven

Leaven is both generated by corruption, and also corrupts the mass with which it is mingled.

Paul also uses leaven to illustrate corruption [1 Cor. 5.6]. Paul encouraged the purging of a sinful man because if his sins remained unpunished they would spread among the group. This statement by Paul cements the traditional meaning associated with leaven of being a corruptive persuasive and a permeating influence.

Of the leaven parable Martin Luther states that:

Our Lord wishes to comfort us with this similitude, and gives us to understand that, when the Gospel, as a piece of new leaven, has once mixed itself with the human race, which is the dough, it will never cease till the end of the world, but will make its way through the whole mass of those who are to be saved, despite of all the gates of Hell. Just as it is impossible for the sourness, which it has once mingled itself with the dough, ever again to be separated from it, because it has changed the nature of the dough, so it is also impossible for Christians to be ever

torn from Christ. For Christ, as a piece of
leaven, is so incorporated with them that
they form with him one body, one mass…
leaven is also the word which renews
men.

Martin Luther assumes that leaven is used merely for its permeating quality and not in relation to corruption. He has interpreted leaven in this parable as depicting faith rather than corruption. The Gospel is the piece of leaven that is incorporated into the dough, which is portrayed as the believers in the human race. Once the leaven has mingled into the dough it can never be separated.

What is the significance of the woman in the leaven parable? Many theologists ask if she is incidental or essential? Most of them see her role as being incidental, as traditionally men's work was sowing and harvesting, whereas, making bread was mainly seen as a woman's work. However, if she was incidental why mention her at all? Jesus could have merely said the leaven was placed in the meal without saying who put it there.

The Bible often uses the figure of a woman to represent kingdoms or cities. In this way, women in the Bible can represent authority and management, especially in the role of hospitality.

The two women represent two covenants. The one whose children are born in slavery is Hagar, and she represents the covenant made at Mount Sinai. Hagar, who stands for Mount Sinai in Arabia, is the figure of the present city of Jerusalem, in slavery with all its people. But the heavenly Jerusalem is free, and she is our mother.

[Gal. 4.26]

The Church is often spoken of as a mother and Catholics often refers to the Mother Church. When women are spoken of in a matriarchal role they are granted greater responsibility. For instance, the Virgin Mary was entrusted with giving birth to the messiah. Therefore, the woman may have been trusted to place the leaven in the meal as she was perceived as being caring and conscientiousness. If the leaven represents Christianity, it appears that she has been entrusted with it.

In practice though women were not often given roles of responsibility. There were no women among the twelve disciples or among the seventy that were commissioned and sent forth. Perhaps because the commissioning was partly undertaken by Paul, the Apostle who did not consider women to be trusted with teaching or matters arising in the church.

Take the teachings that you heard me
proclaim in the presence of many witnesses,
and entrust them with reliable people, who
will be able to teach others.
[2 Tim. 2.2]

Privately women seemed to be held in high regard but publicly they were forced to be silent and not given the same privileges as men. Therefore, if making the bread is considered a domestic and caring role, then the woman could be trusted to place good leaven in the meal.

Women should remain silent in the churches.
They are not allowed to speak, but must be in
submission, as the law says. If they want to

inquire about something, they should ask their
own husbands at home; for it is disgraceful for
a woman to speak in the church.
[1 Cor. 14.34-35]

Similar challenges face women in present day science. Although an equal number of women are being recruited into science at the PhD level, fewer progress to senior positions. Is this because their word and opinion is not taken as seriously as that of their male counterparts?

Women are less likely to receive research grants as they appear to submit fewer applications, and when they do apply for grants they request fewer years of funding and less research money. It seems less likely then, that they will be principal investigators and authors of papers. A researcher's publication list is a very important factor in career progression. Good science is supposed to be impartial and uses evidence-based judgment. Improving female representation at the higher levels could only benefit the image of science.

It could be that women are less competitive than men and are more likely to accept or settle for more menial positions in science or it could be, as Paul's statements illustrate, that they have been socially oppressed into taking more submissive roles. Perhaps, women are simply less likely to be invited to submit journal publications. Certainly if women are given greater opportunity to appear on editorial boards and greater responsibilities within their institutes, this could perhaps improve authorship statistics and subsequent representation of women in senior academic roles in science.

Women in the Bible are often denoted as being the source of corruption. For instance, Eve was persuaded by the serpent into eating fruit from the tree of knowledge. As a consequence of this, the Lord punishes Adam for listening to her but he also punishes Eve with the pain of childbirth and subordination to men.

However, in the parables and teachings of Jesus, he does not appear to demean women. In fact, he is often supportive of them. When a woman adulterer is brought before the Pharisees to be stoned for having broken the Laws of Moses, Jesus defends her by stating whichever one of you has committed no sin throw the first stone [Jn. 8.3-7].

The action of the woman in the parable may be symbolic. She hid the leaven in the meal. If the leaven in this parable is associated with something good she would probably not need to hide it. This could suggest that if the woman was conveying positive doctrines she would boldly speak out. Jesus himself spoke openly unto the world, and his followers perhaps were expected to do the same. He said to them, "Go throughout the whole world and preach the gospel to all mankind."

It is therefore supposed that secret hiding and the spreading of false doctrine are in some way linked. In contrast to the positive message this parable first appears to be conveying, some theologians believe that the woman represents a false messenger, her objective is to introduce a corrupting element into the meal.

On the other hand, Jesus was condemned for his teachings therefore perhaps he is suggesting that his followers spread his message in a concealed way so they do not endure a similar persecution. Alternatively, as previously mentioned, women were discouraged from speaking in the church or in teaching men, so perhaps Jesus is suggesting that women should ignore these restrictions and take a more prominent role in spreading the word of the gospels.

In many translations, there are three measures of meal in the leaven parable. These are thought by many to represent the human race among which the Gospel is working. The number three is mentioned over 500 times in the bible. There are three Magi, Noah had three sons, Lot had three daughters and then, of course, there is the father, son and holy spirit where God is represented in three forms. The three measures of meal in the leaven parable are thought by some to represent faith, hope, and love.

Meanwhile, these three remain: faith, hope,
and love; and the greatest of these is love.
[I Cor. 13.13]

They have also been thought to represent the three elements of human life, body, soul, and spirit, as the meal has been compared to the human race. When Jesus specified three measures, he was trying to portray a figurative meaning that could be interpreted by the people in that space of time. Therefore, it's difficult to know exactly what he was referring too as those influences no longer exist.

In Laws about Sacrifices, in the Old Testament, three measures of meal should be used as part of a sacrifice when a bull is being offered.

When a bull is offered to the Lord as a burnt-offering or as a sacrifice in fulfilment of a vow or as a fellowship-offering, a grain-offering of three measures of flour mixed with two measures of olive oil is to be presented, together with two measures of wine. The smell of the sacrifice is pleasing to the Lord.
[Num. 15.8-10]

There may be an association between the measures of meal and the ceremony of sacrifice. The meal may represent something that is normally sacred and without leaven. The woman may have hidden the leaven into the meal because the teachings of Jesus were in direct conflict to those of the Torah. The Bible always states that no leaven should be offered as a sacrifice to the Lord.

---

*None of the grain offerings which you present
to the Lord shall be made with leaven.*
[Lev. 2.11]

When the woman hid the leaven in the meal she was doing something that was prohibited by God. Therefore, this parable could be deduced to mean the doctrines of Christ are the Bread of Life and must remain pure and uncorrupted. Some propose that the parable of the leaven represents degeneracy in power, a breaking up of divinely ordered fellowship, the corrupting influences of apostasy. The other interpretation is that leaven is, in fact, being used on occasion as a positive element that permeates through the meal as teachings would permeate through society. When Jesus compares his leaven with that of the Pharisees and the Sadducees, although their leaven symbolises corruption his leaven is symbolic of truth.

The leaven parable, therefore, could have two completely different interpretations. On the one hand, leaven could be the corruptive element of sin that a woman hides in the meal which represents the community. Jesus thereby warning how negative words and actions can spread through the community. While, on the other hand, leaven could be seen as the truth being placed in the meal by the woman who represents a believer spreading the good word. With the three measures of meal representing love, faith, and hope. In this context, Jesus is illustrating the opposite, how positive words and actions can also spread through the community. Perhaps it is up to the listener or reader to decide which version they prefer.

Perhaps, the leaven could, in fact, portray two different types of teachings. The woman symbolises the judgment of humanity, she has to judge for herself if the leaven is good or bad and take care in selecting the best kind. If she selects bad leaven, it will fail to grow and the bread will not rise. If she selects good leaven, it will spread through the dough and make higher quality bread. Likewise, if an individual selects bad gospels, it will be very difficult to spread or acquire the understanding needed to enter heaven. If good gospels are selected, the message and knowledge will spread and, therefore, it will be easier to enter the Kingdom of Heaven. The Kingdom of Heaven being a higher state of mind and not a place.

Whereas science tends to be less flexible with only one correct interpretation or truth, religion offers greater flexibility, with several interpretations and many truths.

This could be one source of misunderstanding between the two disciplines.

8

The Blood of Christ

In the age of molecular discovery, it is now known that fermentation is not only responsible for the leavening of bread but is also the principle process in brewing and wine making. It may be worthwhile taking time to consider how wine was perceived in the Biblical era and how its use is viewed in current Christian practices to understand its significance in traditional ceremony.

Present societies now understand that the intoxicating agent of wine is alcohol, a by-product of the yeast fermentation process. It is also understood that unfermented fresh grape juice, or must, is relatively free from alcohol. These principles were not understood by ancient societies, as the knowledge behind the biology of fermentation did not exist. When leaven is used in making bread it is viewed as an impurity and therefore omitted from many sacrificial ceremonies. In contrast wine, which is also produced by a similar fermentation process involving yeast, was not only permitted in sacrifices but was sometimes a principal component.

Wine production occurs naturally in the environment. Grape skins are covered with yeasts and bacteria, mainly members of the yeast family *Saccharomyces*. When grapes are crushed they ferment, especially in warm climates, as yeast fermentation occurs between 20 and 40 °C, with an optimum growth temperature of around 30 °C. The initial fermentation process is in aerobic conditions, the yeast cells are reproducing and producing little alcohol, this continues for a few days. When the yeast stop reproducing they grow anaerobically at a much slower rate and start producing alcohol. When fermentation is complete the resulting wine is racked from the sediment, a substance containing precipitated organic matter and yeast.

In the Biblical era, wine was produced in animal skins or in jars designed specifically for the fermentation process. In the New Testament, the fermentative characteristics of wine were well recognised as is evident in some of the passages. The following parable uses the properties of fermentation to describe how a flexible way of thinking was needed to accept new and fresh ideas.

Nor does anyone pour new wine into used
wineskins, for the skins will burst, the wine will
pour out, and the skins will be ruined. Instead,
new wine is poured into fresh wineskins, and
both will keep in good condition.
[Mt. 9.17; Mk. 2.22; Lk. 5.37]

Wine is not referred to in the Bible as leaven or unleavened although it does feature in sacrifices and rituals. In the Old Testament, it is used in large quantities as part of a daily sacrificial offering that also included animals and unleavened bread [Ex. 29.38-46; Num. 28.1-8]. Wine is also offered on the Sabbath and on the first day of the month, where the quantity varies depending on the type of animal used in the sacrifice. Most notably, wine was offered in the daily sacrifice during the festival of unleavened bread.

The proper wine-offering is two measures of
wine with each bull, one and a half measures
with the ram, and one measure with each
lamb.
[Num. 28.9-15]

It does seem, with respect to sacrifices, that leaven was not associated with wine in the same way that it was associated with bread. Perhaps this is because the process of wine production was not as accessible to the overall population as bread making, therefore it is less likely to be used in domestic ceremonies due to lack of availability. Also, as wine is intoxicating and bread is not, perhaps fermentation in bread was simply thought of as a different process. In addition, wine is rarely associated with food poisoning although it is possible for some microbial toxins to be found in wine.

Generally, if wine becomes contaminated during fermentation it is undrinkable and becomes cloudy, perhaps at this stage, it might have been viewed as impure and corrupt. Pathogenic microbes usually require an optimal pH similar to that found in animals, this is why ethanol with a high pH is normally used in sterilisation. In fact, this could be the reason why alcohol, similarly to salt, is used in these sacrifices, for its ability to sterilise and remove contamination.

As leaven was seen as an impurity that symbolised corruption only unleavened bread is used to celebrate the Passover and to symbolise the body of Christ. The symbolic use of wine in the present Eucharist originated from the words of Jesus at the last supper, which occurred during the feast of unleavened bread.

> *As the disciples were eating, Christ took bread and blessed it, he broke it and shared it among the disciples saying "Take and eat it," he said; "This is my body." He then took the cup, gave thanks to God and passed it to them. "Drink ye all of it; for this is my blood of the covenant, which is shed for many for the remission of sins. But I say unto you, I will not drink henceforth of this fruit of the vine, until that day when I drink it new with you in my Father's kingdom."*
> [Mt. 26.26-29; Mk. 14.22-26; Lk. 22.14-20; 1 Cor. 11.23-25]

During the Passover, Jesus was committed to his fate, he would be condemned to death. He had offended the Pharisees by not obeying the Torah. He persistently associated with people that were unclean, he did not observe strict ritual procedures such as hand washing, he disrupted the traditions of the Temple, and was constantly defamatory towards the policies of the

Pharisees and Sadducees. He was a negative influence that had to be eliminated.

It was the role of the High Priest to sanction and condemn those that had not obeyed the Torah. Jesus knew that he had been betrayed to the high priest by one of his disciples and, in the last supper, took upon himself the role of the Passover sacrificial lamb. It is evident from the Biblical accounts that Jesus was condemned before the Passover as it was against Jewish tradition to execute during a religious festival or on the Sabbath.

Then the chief priests and the elders met together in the palace of Caiaphas, the High Priest, and made plans to arrest Jesus secretly and put him to death. "We must not do it during the festival," they said, "or the people will riot."
[Mt. 26.3-5; Mk. 14.1-2; Lk. 22.1-2; Jn. 11.45-53]

Ironically, this meal is far more poignant as it takes place just before a festival that commemorates the Jews freedom from persecution. Jesus refers to himself as the sacrificial animal used in the traditional ceremony and to the wine as the sacrificial blood but in this instance, the Destroyer did not pass over. Following the meal, Jesus and his disciples retired to the Mount of Olives. Here, Jesus was apprehended by the High Priest's servants who had been led to him by Judas Iscariot, a disloyal disciple. He is brought before the High Priest and the Sanhedrin, the Jewish court that tried those who disobeyed the Torah. He was charged with threatening to destroy the Temple and with blasphemy.

The Sanhedrin delivered Jesus to the Roman procurator, Pontius Pilate, on the grounds that he was claiming to be the King of the Jews and a potential rebel. Roman traditions and Jewish religion would have no doubt been at odds, therefore, Pilate was reluctant to condemn Jesus. Perhaps Pilate shared many viewpoints with Jesus in regard to the Jewish religion and therefore asked the crowd if he should be set free. The crowd responded unfavourably. He was condemned to death by crucifixion, a Roman method of execution. Fearing retribution, the disciples denied their beliefs when interrogated by the Pharisees, but following the death of Jesus they continued preaching his teachings in exile.

The modern Eucharist was established to serve as a reminder of how Jesus gave his life in return for his convictions. In many aspects, this religious ceremony seems to go against the philosophies of Jesus by its ritual connotations and sectarian exclusion. Perhaps serving more as a means of retaining ceremonial sacrifices and symbolic worship of the Pharisees that were rejected in the teachings of Jesus.

In the Eucharist, wine is used to symbolise the blood of Christ but grape juice is sometimes substituted on moral grounds. Many disagree with this principle and regard grape juice as a leavened drink because it has the potential to ferment. Some believe it is impure and it gives rise to objections when it is used symbolically to represent the blood of Christ. Their argument is that wine that has fermented is physically separated from the yeast containing sediment. It is seen as having had the leaven removed and no longer has the potential to ferment, it is predominately thought of as unleavened. Those that follow the doctrines of a Christian religion but strongly object to the moral use of alcohol put forward the argument that the wine used by Christ to represent his blood is a non-alcoholic grape juice.

The Jewish word for leaven is chametz. This is not normally thought to be yeast but naturally fermenting grain, particularly wheat, barley, rye, spelt or oats. In this respect, wine being fermented from grapes would not be considered leavened but beer probably would, as it is likely to contain barley. Sour wine or vinegar was also likely to be considered leavened. The Hebrew for vinegar is chometz meaning sour. This is very similar to the word chametz, which is probably derived from a similar meaning. Vinegar is made by fermenting an alcoholic substance, such as wine, a second time with acetic acid bacteria to convert ethanol into acetic acid. Turning wine into vinegar can be avoided by excluding air from the process as these bacteria are predominately aerobic.

It is likely that wine frequently turned to vinegar in the Biblical era as a consequence of contamination and, therefore, it was thought to simulate corruption in a similar way to leaven. It has been suggested that when the term chometz is used in the Bible it refers to both leaven and vinegar as they are both considered to be sour. The Hebrew word for wine, wasyayin, is derived from the word yaneh meaning to squeeze or press. Sour wine was usually referred to as chometz yayin or chometz yin. It is believed that a type of vinegar or sour wine was offered to Jesus before and after the crucifixion.

*And when they were come unto a place called
Golgotha, that is to say, the place of the skull,
they gave him vinegar to drink mingled with
gall: and when he had tasted thereof, he
would not drink. They crucified him, and
parted his garments, casting lots: that it might
be fulfilled which was spoken by the prophet,
they parted my garments among them, and
upon my vesture did they cast lots. And sitting
down they watched him there; And set up over
his head his accusation written,
THIS IS JESUS THE KING OF THE JEWS.*
[Mt. 27:33-37]

Golgotha was thought to be called the place of the skull
because it was a hill that resembled a skull although it
may have also been called this because it served as a
place for executions. It was located at the entrance of
Jerusalem. Some translations say that Jesus was
offered vinegar while bearing the cross to Golgotha
whereas others say it was wine containing gall. Gall
was often referred to as anything that was bitter, so it
was more likely to be sour wine or vinegar. In his
gospel, Matthew states that this was done to fulfil a
prophecy. The particular prophecy that Matthew refers

to is in the Old Testament. It describes the demeaning manner by which vinegar is offered to quench a thirst. The guards further demean Jesus by removing his clothing in order to share them between themselves. This is also predicted in a prophecy.

When I was hungry they gave me poison.
When I was thirsty they offered me vinegar.
[Ps. 69:21]

Only in the Gospel according to John is it stated that Jesus drank vinegar just before he died while on the cross in order to fulfil the prophecy.

Later, knowing that everything had now been finished, and so that Scripture would be fulfilled, Jesus said, "I am thirsty." A jar of wine vinegar was there, so they soaked a sponge in it, put the sponge on a stalk of the hyssop plant, and lifted it to Jesus's lips. When he had received the drink, Jesus said, "It is finished." With that, he bowed his head and gave up his spirit.
[Jn. 19:28-30]

It would seem that vinegar could be more likely to be viewed as a leavened drink than either wine or grape juice as it was synonymous with a corruptive influence in the same way as leaven.

In the New Testament messages are communicated so that they are accessible to those they are expected to influence. If there were a spiritual objection to drinking alcoholic wine than surely these ideals would be put forward in Biblical teachings? And, if this were the case, why not use water as the pure and sinless drink in the last supper? The Hebrews seemed to have an innate sense of disease prevention. In sacrifices they used animals that had no blemishes, they ostracised those individuals that were viewed as unclean, they removed potential microbial contaminants, such as leaven, from food. Water in that era would have been the source of many contaminants and likely to contain as many microbes as leaven, therefore wine would be less likely to cause disease. Water could, therefore, be interpreted, as a disease-causing agent in contrast wine would have been associated with disease prevention. As the processes behind diseases were unknown they were attributed to acts of retribution by angry deities. In the disease-ridden era of the Biblical age, a gift of wine to the Lord would perhaps have been perceived as more suitable than a gift of water.

The Leaven

The inexplicable nature by which grape juice turned into wine was at one time the subject of much speculation. In 15th century England, leaven used in brewing was known as *barm*. In the Brewers Book of Norwich, written in the 15th century, the barm is referred to as *goddisgoode* because it was thought to be provided by God's blessing. In the absence of understanding, God was invoked as the great provider. It seems inevitable that wine could be associated with miracles, given the mystery surrounding its existence.

Perhaps one of the most controversial miracles that Jesus performed was at a wedding in Cana in front of his disciples. It was claimed by John to be his first miracle.

There was a wedding in the town of Cana in Galilee. Jesus's mother was there, and Jesus and his disciples had also been invited to the wedding. When the wine had been given out, Jesus's mother said to him, "They have no wine left."
"You must not tell me what to do." Jesus replied, "My time has not yet come."
Jesus's mother then told the servants, "Do whatever he tells you."

The Jews have rules about ritual washing, and
for this purpose six stone water jars were
there, each one large enough to hold about a
hundred litres. Jesus said to the servants, "Fill
these jars with water." They filled them to the
brim, and then he told them, "Now draw some
water out and take it to the man in charge of
the feast." They took him the water, which
now had turned to wine, and he tasted it. He
did not know where the wine had come from
but of course the servants who had drawn the
water knew; so he called the bridegroom and
said to him, "Everyone else serves the good
wine first, and after the guests have had
plenty to drink, he serves the ordinary wine.
But you have kept the best wine until now!"
[Jn. 2.1-12]

This miracle seems to differ from others in that its main purpose is to demonstrate that he had a divine gift that distinguished him from ordinary people. The wedding guests have already had their fill of wine and yet desire more. Here a miracle that does not address any spiritual issues other than to appease Jesus's mother, who seems distressed at this lapse in hospitality. This story is only mentioned in the gospel according to John.

The first three gospels in the New Testament by Matthew, Mark and Luke show many common elements and have verbal similarity they are thought to be the more accurate records of Jesus's ministry and have been written about the same time. Collectively they are known as the Synoptics as they share a common perspective. The fourth gospel according to John tends to show Jesus as a messiah and, therefore, may exaggerate some of the miracles performed. The objective of this first miracle according to John was for Jesus to manifest his glory and for the disciples to believe in him.

This, the first of his signs, Jesus did at Cana in Galilee, and manifested his glory; and his disciples believed in him.
[Jn. 2.11]

Through the disciples, the miracles were used to persuade others to follow Jesus's teachings. The miraculous transformation of water into wine at the wedding of Cana is taken as a sign that Jesus sanctioned alcoholic beverages and their consumption was viewed as a socially accepted activity. Simultaneously, Jesus acknowledged the sanctity of marriage. This miracle was seen by some as an indication that Jesus's actions would enrich the lives of the community by benefiting their social needs. According to John, the news of this miracle filtered through the local community. When Jesus later visited Cana he was greeted by a government official who thought his son was critically ill and dying.

Jesus said to him, "None of you will ever believe unless you see miracles and wonders."
"Sir," replied the official, "Come with me before my child dies."
Jesus said to him, "Go your son will live!"
The man believed Jesus's words and went. On his way home his servants met him with the news, "Your boy is going to live!"
[Jn. 4.46-51]

This second miracle symbolised that not only could Jesus benefit the social needs of the community he could also protect them from harm. Jesus had the power to heal and to create. These two examples of miracles have a common denominator they both exploit social views of uncertainty. The first miracle exploits the views of uncertainty surrounding wine fermentation the second exploits the uncertainty of fear associated with disease and death. Two situations that, although are beyond the control of human intervention, could be resolved by divine interaction.

The words for wine used in the New Testament are *oinos*, a Greek term for completely fermented wine, and *gleukos*, used to denote new or sweet wine with less alcohol content. Gleukos as a reference to wine that has been drunk is only mentioned one time in the New Testament [Acts 2.13]. In the context of this passage, the apostles were behaving in an unusual way because they were full of the Holy Spirit. Onlooker's accused them of behaving as if they were drunk on gleukos because of how their behaviour had changed with no alcohol being present.

Biblical society viewed wine in a similar way to how it is currently perceived. They were aware that over indulgence could be harmful but generally it was socially accepted. The New Testament attempts to rescue individuals from a drunken abyss by suggesting that they should be filled with a different kind of spirit.

Do not get drunk with wine, which will only
ruin you; instead, be filled with the Spirit.
[Eph. 5.18]

In a comparatively brutal manner, the Old Testament illustrates and blatantly condemns the consequences of intoxication.

The Lord God said to me, "Jeremiah, tell the
people of Israel that every wine-jar should be
filled with wine. They will answer that they
know every wine-jar should be filled with wine.
Then tell them that I, the Lord, am going to fill
the people in the land with wine until they are
drunk: the kings, who are David's descendants,
the priests, the prophets, and all the people of
Jerusalem. Then I will smash them like jars
against one another, old and young alike. No
pity, compassion, or mercy will stop me from
killing them."
[Jer. 13.12-15]

The Roman's took wine very seriously, to the extent that they even had a deity assigned to it, *Bacchus*.

Even so, some social groups were discouraged from drinking alcohol. For instance, women, in the early day of the Republic, were forbidden from drinking ordinary wine but were permitted to drink those with low alcohol content. There were a number of ways that the alcoholic content of wine could be reduced. Fermentation could be inhibited by increasing sugar content. The Romans called this beverage *defrutum*. Grape juice with enough sweetness to remain unfermented can be made by pressing dried grapes. Pliny refers to a raisin-wine, made from grapes dried to half their weight. Roman women also drank a wine alternative made of raisins called *passum*.

Another method to reduce wine alcoholic content was to prevent the yeast from growing. Vinous fermentation occurs only within a certain temperature range, the lower limit is about 15 °C. When cooled wine was allowed to sit undisturbed, the clear juice could be removed from the sediment and would remain unfermented. This would perhaps have been difficult to achieve in hot climates. Another method of making a none alcoholic wine was by adding salt, a process favoured by the Greeks and described by several classical authors (Cato, Columella, and Pliny), this method was also used to preserve the must. Alcohol evaporates at around 100 °C, so could be physically removed from the wine by heating. Pliny describes another drink called *adynamon*, made by adding water to wine and boiling the mixture until the quantity was considerably reduced. This provided a fortifying drink for invalids.

It is believed that the Hebrews were also familiar with preserving wine by boiling down grape juice to a thick syrup like molasses. The boiling process would also remove any microbial contaminants from the grapes. The syrup would be diluted with water as a drink or added to wine must. Some of the Biblical references to honey debash could be referring to a sweet grape syrup. The Hebrew debash is similar to Arabic dibs, a sweet syrup made by boiling down the juice of grapes, raisins or dates.

In moderate use the social impact of yeast is beneficial, but alcoholism is a modern scourge of the 21st century, overshadowing the problems caused by microbial diseases. In a recent study conducted by the World Health Organisation, the long-term health burden of alcohol-related disease surpasses smoking and malnutrition. The countries that produce the highest quota of alcohol last century were the USA (beer), China (spirits), and France (wine). The leading exporter of alcohol was Great Britain, which exports nearly twice as much as France, in second place. The total consumption of alcohol increased in Great Britain between the 1970s and 1990s, while in France it decreased. However, in the 1990s, the French were still more likely to consume more alcohol per capita than the British, 14% compared to 9%.

In general, the Bible seems to portray the message that drinking wine is an acceptable part of everyday life, but its increased accessibility by modern preservation and production methods seem to have created new social challenges. Leaven is not characterised in wine in the same way that is in bread. Some argue that the leaven is removed once fermentation has ceased so although fermenting must, contaminated wine, and vinegar could be viewed as being leavened, wine itself is pure.

9

Leaven in a Molecular Era

Not only does yeast now serve as one of the most important organisms throughout domestic history, in recent years it has also substantially contributed to biological research. The numerous molecular techniques that have evolved in yeast have allowed it to make an important contribution to a number of areas in science. Through studying various types of yeast and other microbes, scientists now know a great deal about the molecular processes involved in cell division, rapid evolution, and disease.

Large-scale experiments involving computers, robotics, and molecular techniques, such as polymerase chain reaction (PCR), to amplify genes, and DNA microarrays, that arrange hundreds of these amplified genes onto chips, have generated such a large amount of data that new scientific disciplines, e.g., genomics, transcriptomics, etc., have evolved in order to process it all into meaningful results.

Fortunately, individuals with skin diseases are no longer thought of as unclean and are normally treated within the community. Scientists have a greater understanding of disease management and although quarantine and hygiene are still practiced, they are now carried out within certain timescales in order to reduce disease transmission. In the majority of cases, people are not ostracised when they are infected by disease, although fears and anxieties can still be generated through sensational media coverage. Nevertheless, even in this molecular age, some transmissible diseases are still associated with sins of the flesh and can lead to social ostracisation.

There are still many diseases that generate fear because they are untreatable. Some of these have evolved through human activities, such as bovine spongiform encephalopathy (BSE) which gives rise to a human form of spongiform encephalopathy called variant Creutzfeldt–Jakob disease (vCJD). The causative agent of BSE is a defective version of a protein called a prion that is similar to one found in the brains of sheep with scrapies. The prion protein is transmitted horizontally and causes disease through disrupting the normal function of the native protein. Studying the molecular mechanisms by which proteins change conformation to become prions in yeast has led to a greater understanding in the pathology of this disease. Many other human diseases, especially cancers, can be researched by studying molecular processes first in yeast.

Cancers arise when cells begin to divide abnormally due to mutations. Cancer research investigates the mechanisms that encourage these mutations to arise. The mechanism of cell division is often studied in the fission yeast, *Schizosaccharomyces pombe.* Unlike *Saccharomyces cerevisiae*, which divides by budding, *S. pombe* divides symmetrically in a similar way to human cells. Fission yeast originates from Africa, where it is found growing on banana skins and is used to ferment beer.

Through researching cell division, scientists have reached many milestones in the mechanisms that have caused various cancers leading to greatly improved clinical treatments. Work in yeast genetics has greatly contributed to our understanding of cell cycle research as researchers have found similar cell division genes in human genomes. This led to the award of a Nobel prize in 2001 to three scientists who led pioneering work in this area. The three scientists were: Paul Nurse, for his work in *S. pombe* and human model systems; Leland

Hartwell, for his work in *S. cerevisiae*; and Tim Hunt who used sea urchins as a model system.

In addition to investigating diseases, yeast is also used as a model system to research ageing. *Saccharomyces* cells can divide by budding a number of times but the new bud is always physiologically younger than the mother cell. Each cell produces about thirty buds, depending on the environmental conditions and other factors, and then it dies. About thirty genes in yeast have already been found to be involved in ageing. The main factors seem to be related to metabolic capacity, resistance to stress, gene dysregulation, and genetic stability. Encountering certain environments that would overload any of these factors would also affect longevity. For instance, excessive oxidative damage or radioactivity would lead to a high level of mutations that will reduce the number of times that a cell can bud. Excessive oxidation is associated with the consumption of calories, so caloric restriction should result in increased longevity. This has been demonstrated in yeast, limiting the amount of nutrients and carbohydrates available in growth medium leads to a longer generation time and lifespan.

The development of yeast molecular biology can literally be used to assess the impact that the application of scientific research has had on 21st century society. Scientific researchers often describe yeast as the workhorse of eukaryotic molecular biology with many laboratories devoted to studying this single-celled organism, as much of the information derived from it can be equally applied to the study of human cells.

Most modern laboratory strains of yeast originate from one particular *S. cerevisiae* strain, EM93, isolated from dried figs in Merced, California in the 1930s by Emil Mrak. This strain turned out to be heterothallic, meaning that cells existed as two types of sterile haploids, with a single copy of each gene, that when fused together formed a fertile diploid that could perform meiosis in a similar way to that seen in human cells. Up until this point, most strains studied were homothallic, this meant that all haploid cells were of the same mating type and capable of fusing together to form a fertile cell known as a zygote. The emergence of a heterothallic strain meant that the genetic stability of a culture could be placed under greater control, as it would remain haploid until a different haploid strain was introduced and then, through the production of mating pheromones followed by cell fusion, a diploid cell could be created.

Another reason why yeast is used as a molecular model system alongside other well-known microbes, such as *Escherichia coli*, is because it is a eukaryote. *E. coli* and other bacteria are prokaryotes, in contrast to eukaryotes they only have one chromosome housed in a cell without a nucleus. In yeast cells, DNA is packaged in chromosomes stored in a nucleus in a similar way as in human cells. Yeast has 16 individual chromosomes compared to 23 in humans. Surprisingly, there are only four chromosomes in the multi-celled fruit fly *Drosophila*, another model organism used for biological research. Yeast also has the advantage of being able to grow just as happily with one set of chromosomes, in haploid cells, as with two or more sets of chromosomes, diploid and polyploid, respectively. Additionally, as yeast is a single-celled organism without the complexity of cellular differentiation, by which organs are formed, it can be used to study the cell cycle at a fundamental level.

Yeast is used to research both mitosis and meiosis. Many mutations that cause human disease are introduced during meiosis. Following cell fusion or mating, two haploid cells form diploids, which can produce four individual haploid cells in the form of ascospores known collectively as a tetrad. After microscopic dissection of the tetrads into individual ascospores, researchers can study recessive mutations and the complicated exchange of genetic material during meiosis by counting the numbers of surviving progeny. For instance, the distance between two molecular markers on a chromosome can be measured by the number of ascospores that carry either of the markers.

The information derived from yeast studies aids the study of genes involved in tissue development and cell differentiation in higher eukaryotes, such as *Drosophila*. Adding to all these factors many of the biochemical and cellular functions in yeast are conserved in human cells. Yeast, therefore, is a simple and practical system to study the mechanism of human cell division.

So, why has yeast become such a popular organism to study molecular biology and why has this microbe been chosen in favour of others microorganisms? Firstly, *Saccharomyces* is non-pathogenic and does not present a threat to human safety. Therefore, laboratory workers do not require expensive protective equipment to practice research. *Saccharomyces* is also easy to contain as it is not usually airborne unless transported involuntarily by animals and insects. It can also be grown easily and only requires a suitable carbon source, nutrients and appropriate physical conditions to continue multiplying. Additionally, these requirements can also be used to control the rate of cell division, for instance, by altering temperature or by creating metabolic mutants that can only grow in certain media.

Mutants are generated either through using a mutagen or by manipulating DNA through genetic engineering. Genes involved in yeast metabolism can be mutated and then used as molecular markers. For instance, if the genes for the requirement of an essential amino acid are defective then the yeast will not grow without that amino acid added to its immediate environment. If the defective gene is artificially replaced by a functional one, then the yeast cell will be able to continue growing without the need for that particular amino acid. Armed with this knowledge, researchers are able to introduce fragments of DNA fused to these markers or reporter genes. If the yeast is able to grow without the selected amino acid this means that the DNA of interest to the researcher has been successfully introduced into the cell. This approach has led to the characterisation of countless genes and proteins in yeast and from other organisms.

The last century saw a molecular enlightenment which accelerated our understanding of the natural world; yeast was a key component of that movement. In the

1950s, molecular biologists constructed a *S. cerevisiae* strain containing biochemical markers (antibiotic resistance or amino acid selection) known as S288C from the previously mentioned fig strain, EM93.

It was soon discovered that self-replicating elements of DNA found in bacteria called plasmids, could also be made to function in yeast. Yeast cells multiply rapidly and the overall effect of a mutation in a certain gene can be measured biochemically or by observation under the microscope. If DNA fused to reporter genes is inserted into the self-replicating plasmid from bacteria and then introduced into the yeast cell, it can be propagated and then extracted. This technique is called cloning as it replicates an identical copy of a gene, it has been used to mass-produce proteins and vaccines.

Cloning was used in yeast as early as 1980 to produce Hepatitis B vaccine. Since then it has produced a multitude of proteins and vaccines including insulin, growth hormone, haemoglobin, oestrogen receptor, and interferons.

Cloning can also take place in bacteria such as *E. coli*, these cells divide faster than mammalian cells but are a lot smaller, so there is a limit in the size of the protein that can be cloned. As a consequence of this other cells types are now also used for cloning such as those derived from mammals, insects, and viruses. Cloning provides an extremely economical way to reproduce human proteins. They replace the need for animal production and reduce the risk of transferring unwanted diseases, such as CJD from growth factor.

Although great advances have been made in cloning, the systems are still not perfect and have their limitations according to the type of protein that can be cloned. Some proteins are toxic to the cell, and the introduction of unwanted mutations occurs far more frequently when selection is not acting on the protein. Most organisms, including bacteria, have their own DNA repair systems that detect mutations. Foreign DNA has a higher chance of retaining mutations in a host cell as it is not detected through normal cell function, a problem to biotechnology that is successfully addressed in natural systems by selection. The use of these systems with these limitations causes uncertainty and increases risk factors, subjects which are discussed later. It seems yeast occasionally retains its Biblical ability to behave in a corrupt way.

Following the heady days of protein engineering, yeast laboratories, through intra-science communication, completed an enormous challenge - the first fully sequenced eukaryotic genome. This was achieved using *S. cerevisiae* strain S288C and relatively archaic apparatus compared to the robotic systems used to decode the human genome. Eventually, over 6,000 genes were unravelled from the yeast nucleus. The yeast genome is 200 times smaller than the human genome but almost four times larger than that of *E. coli*. This achievement marked a milestone in biological history.

Yeast biologists did not stop at just sequencing the genome. In a remarkable example of inter-organisation collaboration, nominated laboratories began deleting single genes from individual yeast cells. Through using PCR, a marker/reporter gene flanked by target DNA was amplified and then inserted into the yeast cell. Through using the cells natural recombination, the genomic gene was replaced with the introduced non-homologous marker/reporter gene. Biochemical tests were then carried out on the mutant yeast strains to uncover the functional analysis of hundreds of different gene products. This work elucidated many gene functions and undoubtedly contributed to the discovery of many analogous human genes. This information was collated into several databases to provide a plethora of data available for bioinformatics across the internet. Genes placed on microarray slides and subjected to various environmental conditions and variations of DNA recombination techniques have increased the quantity of this information, enabling researchers to compose complicated hypotheses and uncover new cell processes without even entering a laboratory.

Many would expect yeast's contribution to scientific research to stop at this point. Exhausted by constant, investigative probing. In contrast, the yeast story continues. It has also been used as a vehicle to investigate protein interactions first with native yeast proteins and then later with proteins from other organisms. Genes can be fused to protein tags, then introduced into yeast cells where reporter genes within the cell can detect if the proteins, produced from the introduced DNA, interact with each other. This procedure is known as the yeast 2-hybrid (Y2H) technique. Several variations of Y2H exist and it also has been modified to be used in cell cultures from other organisms.

These techniques, in a rudimentary way, can also be used to evaluate post-translational modifications in proteins, to see how gene products are modified by the cell. Compared to the amount of gene sequencing data available the amount of protein interaction data is still fairly incomplete with the function of many gene products still unknown, but this is rapidly changing.

Yeast has also made a valuable impact in evolutionary biology as it has allowed the mechanisms of evolution to be scrutinised at the molecular level and over short time-scales. In evolutionary terms, fungi, including yeasts, precede mammals and other bilaterians. Bilaterians possess a left and right symmetry of body plan. The two predominate groups, deuterostomes and protostomes, differ from one another in skeletal development. They are believed to have separated in an early stage of evolution estimated to be 670 million years ago.

Humans are likely to have diverged from apes only four to five million years ago. Plants and fungi are thought to have moved from water to land together, the earliest fossils of fungi are in Precambrian rocks dating back 900 million years. Comparing conserved DNA motifs between species of yeasts, allows geneticists to estimate the evolution rate of proteins. Yeast can be compared with other yeasts and then with other model organisms, such as, nematodes or fruit flies. Comparative genomics evaluates the evolution of certain proteins and the processes and complicated pathways that they participate in.

Fungal species are susceptible to disease and parasites that they control by producing antibiotics, such as, penicillin. In fact, the microbial world is full of toxins secreted by bacteria and fungi, many are used as insecticides and other biological control agents. Yeast can also be used to study antibiotic resistance. Resistance to antibiotics and other stresses in yeast is often called rapid evolution. As, yeast cells evolve rapidly to overcome environmental challenge, they provide a means to study the mechanisms of evolution. In addition, the yeast cell susceptibility to mutagens make it an ideal organism to study the effects of mutagenesis and adaptation.

Yeast, therefore, provides a molecular tool to study cell biology and a model system that can add to our knowledge of evolution. In contrast to yeast in the Biblical era, the Molecular era now knows a great deal about this organism. As well as greatly improving disease management, advances in genetics have led to new arguments surrounding the creation of living things, especially in respect to evolution and cloning. Even though yeast exists only as a simple single-celled organism that thousands of researchers have been studying intensely for centuries, a lot still remains to be discovered.

Life on earth has evolved over millions of years through a complex network of processes that will take many years to unravel. Whether the molecular information we have derived from yeast is comparable to the corrupt leaven of the Pharisees or a leaven that represents the kingdom of heaven, has yet to be established. However, the leaven that the woman hid in the meal is very similar to how research is dispersed through science. Good research, in a similar way to good leaven, will lead to a growth of knowledge, whereas, bad research will cease to grow.

10

Beware the Leaven

Whereas religion uses faith to dispel anxiety caused through uncertainty, science calls upon facts to achieve the same purpose. In the Old Testament, leaven was omitted from sacrifices in order to increase the purity of offerings thereby eliminating the chance of offending a deity. The fact a sacrificial offering was taking place at all accentuates the extent that fear manifested through uncertainty. Individuals would go to enormous lengths and effort in order to eliminate risk.

Through scientific evaluation, most of us now rely on facts rather than sacrificial offerings to protect us from disease and other catastrophes. For instance, we are aware that food and drug administrative laws exist to protect consumer's interests. Although, this still does not entirely eliminate uncertainty, as trust in the regulators also relies heavily on faith. The public must rely on scientists, and other professionals, to have obtained experimental evidence according to the ethics of the law.

In pharmaceutical development, research occurs initially within a controlled laboratory environment until a new drug is released on to the market, at this point there is no barrier between science and the public. Whenever we take a prescribed drug we assume that it is going to be beneficial or, if this were not the case, the consequences to the prescriber would be so detrimental that the risk of malicious activity would be fairly small. Occasionally, drugs are released without all the necessary research being completed, sometimes with catastrophic consequences. A classic example of this was the thalidomide scandal in the 1960s.

Thalidomide still evokes images of human suffering, a reputation it earned because of the deformities it inflicted onto unborn children. Thalidomide was given to pregnant women to assist sleep and prevent morning sickness. The drug interfered with foetal development to produce deformities that included missing or abnormal limbs, spinal defects, cleft palates, and the abnormal formation of many vital organs. Forty percent of cases lead to mortality during or shortly after birth.

The German pharmaceutical company, Chemie Grünenthal, first marketed thalidomide in 1957 as a hypnotic to induce deep sleep without producing the side effects associated with other barbiturates available at the time. A year previously, in 1956, the research of Wilhelm Kunz found it depressed the nervous systems of animals without fatalities. It was considered to be a safe alternative to contemporary medicines because its low toxicity could prevent an accidental or intentional overdose.

A marked increase in deformities among newborns of patients taking the drug caused physicians to demand thalidomide be withdrawn from the world market in 1961. Astonishingly within a relatively short span of time between the first appearance of the drug and its subsequent withdrawal it had adversely affected the lives of more than 10,000 individuals.

Controversy surrounded the way in which the drug companies had marketed and produced thalidomide. It was argued that the disaster could have been averted if correct scientific procedures and protection were followed. Not until after the drug was removed from the market were extensive reproductive tests in animals carried out. Normally the pharmaceutical company would carry out and publish extensive reproductive studies in animals. Yet in the case of thalidomide, a drug that was being prescribed to 'pregnant women', similar studies had not been undertaken. Controversies, such as this, reflect the ambivalent relationship between science and other social institutions such as the media, regulatory systems, and commerce.

The media played an important role in the thalidomide controversy, contributing to legal history when the Sunday Times won an important case against Her Majesty's Attorney-General in the European Court of Human Rights. The article that evoked this response exposed the shortcomings of the Distillers Company (Biochemicals) Limited, that marketed the drug as Distaval. Lack of experimental studies prevented the drug from reaching the American market as the chronic toxicity data was incomplete. The Food and Drug Administration (FDA) criticised the lack of long-term scientific studies and was concerned about the evidence that the pharmaceutical company was withholding, they were particularly apprehensive about the drugs reported side effect of peripheral neuritis.

The thalidomide controversy immediately led to tighter controls on the introduction of new drugs in the UK. An independent Committee on Safety of Drugs was formed and eventually a Medicines Act was established in 1968, followed by the formation of a Committee on

Safety of Medicines in 1971. In comparison, the USA had the Pure Food and Drugs Act in place as early as 1906. Evaluation to ensure drug safety was established by The Federal Food, Drug and Cosmetic Act in 1938 governed by the FDA that was founded in 1931. The FDA demanded scientific evidence to evaluate thalidomide, it was not available and therefore the drug was not granted a license to be marketed. If the same procedures had been followed in other countries the number of affected individuals may have been substantially reduced.

The thalidomide controversy drew the public's attention towards the morality of pharmaceutical companies and many aspects of scientific research. Tensions existed between public health issues and economic priorities. Pharmaceutical companies exploited scientific discovery for financial gain that appeared to obliterate human compassion, with individuals passing responsibility so that nobody was eventually held to account. Not only were the victims facing the physical and mental hardship of overcoming severe disabilities, but they also faced a long fight to receive compensation.

The science of actually manufacturing the drug was not fully responsible for the thalidomide controversy. If thalidomide occurred naturally, say as a plant extract, and was not synthesized in a lab these problems would still have existed. It was the marketing of the substance as a drug that caused the problems. The lack of scientific evidence that ensured the drug was safe. The public learnt that scientific evidence could be manipulated by financial gain. Controversies such as this cause people to lose faith in science. In this case essential research was not presented in order to release the drug on to the market and the authorities failed to regulate the pharmaceutical company concerned or compensate the victims.

Advances in technology can bring significant medical or agricultural benefits but they are also exploited for commercial gain, sometimes with a disregard to any negative impact they may have. In general, the introduction of new drugs has benefited human health. The early 1900s witnessed many great advances in chemical therapy. Ehrlich developed a treatment for syphilis in 1907; Fleming discovered the first antibiotic, penicillin in 1920; Domagk found that sulphanilamide could cure septicaemia in 1932. Life expectancy has improved as a consequence of these drugs, in 1935 there were 3,690 deaths in Britain from scarlet fever and diphtheria compared to just one death in 1970. Thalidomide was introduced hot on the heels of these discoveries, when huge profits could be made from a new drug. Especially in Britain, where there was less regulation of pharmaceuticals.

Ironically, because of extensive screening and restrictions now in place to prevent a reoccurrence of an episode like the thalidomide tragedy, it has become increasingly difficult to bring new drugs on to the market. Therefore, re-evaluating harmful or ineffective drugs for other uses has become an ever increasing trend with pharmaceutical companies. The mechanism behind the teratogenicity of thalidomide has still not been established yet this drug is currently being researched as a possible therapeutic agent for other diseases, such as cancer. Celegene are currently remarketing thalidomide to treat the symptoms of leprosy. It is also effective in the treatment of some myeloma.

In the documentation that is produced to promote the thalidomide (Thalomid: Balancing the benefits and the risks), it is admitted that birth defects still occur in countries where controls and monitoring plans have been inadequate. This brings about another form of controversy, where there is tension between the victim of a disease and a social goal to obliterate thalidomide.

The Sunday Times article, which the High Court tried to ban, ended with an emotional statement.

Many of the main characters who figured in this narrative are now in other employment, thalidomide is only a painful memory. Of the original cast of the tragedy only the victims still occupy the stage.
[The Sunday Times, 1977]

Thalidomide still retains the ability to promote controversy and to raise issues by the public concerned with the morality of scientists under the influence of commercial gain.

So metaphorically speaking, whose leaven should modern society now be aware of? There are a number of candidates that can permeate a corruptive influence in science, including the media, researchers, commercial companies (pharmaceutical, agricultural, etc.), academics, and politicians. The majority of science is fairly mundane and straightforward, occurring without grabbing news headlines but occasionally something attracts media attention and becomes headline news. On several occasions in the past, trust in science has been challenged by controversy.

The introduction of genetically manipulated or modified (GM) products into the environment before all research has been effectively collated and publicly distributed, can manifest into controversy. The public may be justifiably cautious of GM food, if there is no benefit in eating it, then why take a risk? An experiment that is restricted to a laboratory can be controlled and environmental conditions can be manipulated. Once an experiment leaves the controlled confinement of a laboratory it is at the mercy of a number of influencing and unpredictable factors, including commercial profit. It was exactly these factors, combined with the lack of governmental policy, that contributed to thalidomide entering the marketplace.

Following the impact of the BSE crisis, Britain is very cautious about the introduction of GM crops. In the United States GM crops are fairly prevalent, accounting for over 90% of all soybean and rapeseed production. Most GM crops are resistant to glyphosate herbicides, but there are also GM pest-resistant crops. In the US 90% of cotton production is GM for pest resistance. In 2010, 10% of the worlds crop production was GM most of which is grown in the US. Invariably, most of us are unwittingly eating some form of GM food.

In food production, GM yeasts are available for wine production and in baking. In the UK two products have been approved for commercial use. One in baking to reduce leavening time and the other in brewing to produce low calorie beer. At the time of writing, no GM yeast were used in EU countries. In the US and Canada, a GM yeast called ML01 is used to improve the taste and colour of wine and to reduce the production of histamines. Additionally, a GM yeast that has been produced to reduce the production of ethyl carbamate, a compound with carcinogenic properties, has been generally recognised as safe for use in the US. Most GM yeasts have non-food usage, they are grown in controlled environments to produce pharmaceuticals, chemical compounds, and enzymes. They are also increasingly being developed to produce biofuels.

Very little research has been carried out on the safety of consuming GM products. Very often a spokesperson will be quoted as saying that there is no research to imply that GM crops are harmful, however, more often than not, there is no research to say that they are not harmful either. The consequence of eating pesticide containing GM crops, such as maize, is largely unknown. Perhaps scientists are wary about publishing research because they face academic peer pressure. Future grants may rely on a product not being tested unfit for human consumption.

In 1998, researcher Árpád Pusztai reported changes in the intestines of rats who had eaten a GM potato that could confer pest resistance by producing a lectin found in snowdrops. Other scientists in the field argued there were insufficient controls to determine that the toxin, and not the potatoes, was detrimental to the rats. As a consequence of his research, Pusztai was suspended from his post and his contract was not renewed. This incident, known as the Pusztai affair, highlights the extent at which scientists and their research are influenced by external factors, outside their control.

Although pharmaceuticals, in general, provide us with a lifespan and standard of living that far surpass that in the Biblical era, controversies can still occur. Media that implicates the use of a new drug in compromising the development of children, can rapidly ignite public concern. These issues cause outrage because they effect the vulnerable, those individuals that have no choice but to rely on the judgment of others. Who in turn must have faith in pharmaceutical companies and the bodies that govern them. The problem with governing bodies is that individuals within them are usually specialists who can be influenced by other interests or individuals. Also, responsibility is spread within a group, there is no single individual who has to take account of any misdemeanour.

In 1988, the measles, mumps and rubella (MMR) vaccine was introduced into Britain. ten years later Dr Andrew Wakefield and his colleagues at the Royal Free Hospital, London, found that a bowel condition could be associated with autism in a study group of 12 children. Traces of measles virus were discovered in the intestines of these children but the data established an association rather than an effect. There was no evidence that the single measles vaccination would not behave in the same way, however, the incidence of autism has increased in recent years. Many parents were concerned that the triple vaccine was harming their children.

Autism can be related to other conditions and consequently it is fairly difficult for the disease to be linked to a particular vaccine. It is virtually impossible to prove a cause and effect. Even if in reality some individuals were more susceptible to autism after receiving the vaccine, it would be difficult to directly link the vaccine to the disease.

Astonishingly, even though Wakefield did not specifically state that the MMR vaccine caused autism, the UK government, pharmaceutical companies, and medical research bodies exploded into a retaliation that was completely disproportionate. He was forced to resign, accused of professional misconduct, struck off the medical register and several of his papers have been retracted by journals. This is quite disturbing for researchers, as it prevents them and scientific journals from publishing research that may attract controversy.

If there was an element of uncertainty in vaccinating individuals then why not stop the vaccines and go back to the old methods of vaccination or better still, offer both choices? After all, the old methods seemed to be fairly effective in preventing the spread of measles, mumps and rubella. Additionally, if both methods were available then it would be possible to compare the rates of autism in a greater population and, therefore, the epidemiology would be clearer. Apart from the obvious financial considerations, the answer perhaps lies in the statistical way that governing bodies evaluate risk. If the risk of becoming autistic is below a significant level than it would be argued there is no risk in having the

MMR vaccination, even if that risk is double that of the single vaccinations. Several members of the public found that a potential risk, however small, of being permanently harmed by the MMR vaccine was greater than the risk of temporarily contracting measles and, therefore, chose not to vaccinate their children. In many areas, only two in three children were vaccinated leading to concern that a measles epidemic could occur and result in disabilities or even fatalities.

Media reporting had originally alerted parents to the potential of MMR in causing autism. In a survey conducted by the Economic and Social Research Council the majority of people discovered the controversy surrounding MMR by watching the television. In fact, television is the most important means of communicating science controversy with science journals providing the source information. So, could it be possible that sensational media coverage was responsible for the MMR controversy and that Wakefield was merely used as a scapegoat to counteract this?

Interestingly, Japan banned the use of the MMR vaccination in 1993, when at one time it was compulsory. They carried out a series of investigations after finding that the incidence of non-viral meningitis and other adverse reactions had reached record levels. An analysis of the vaccination, over a three-month period, showed one in every 900 children experienced adverse reactions, this was 2,000 times higher than the expected rate of one child in every 100,000 to 200,000. Traces of the vaccine were found in spinal fluid. The incidents of autism were not recorded, but it would seem that the vaccine had the capability of causing adverse neurological effects.

A similar concern to the MMR vaccination, that did not attract as much media attention, was the discovery that the anthrax vaccination, given to armed service personnel in order to protect them from Biological weapons, could perhaps cause miscarriages and premature births. If this problem had affected a greater percentage of the public, it would naturally have been given more media attention. As it happens, this could never have occurred as the product was not licensed for general release. Therefore, the only people who were likely to be at risk were also at greatest risk of encountering anthrax spores.

Controversially, despite originally being an animal vaccine, the American and British governments claimed that animal reproductive studies on the anthrax vaccine had not been carried out. The incidence of birth defects, mainly associated with premature births, and miscarriages in the infants of American and British troops has increased since the anthrax vaccination program began. Initially, in the early 1950s, when the vaccination was licensed for use on humans, only male service personnel would have been vaccinated. In the 1990s, when troops were vaccinated in order to go to the Gulf, both men and women received the vaccine.

The vaccine had already earned a bad reputation among service personnel for causing ill health. In order to acquire complete immunity over three shots were required and many of those that received the vaccine voluntarily, failed to complete the course with some remaining ill and developing other symptoms. The vaccine was thought to be responsible for Gulf war syndrome, as countries that did not vaccinate their troops did not have cases of Gulf war syndrome either.

Governments and pharmaceutical companies have refused to acknowledge accusations against the anthrax vaccine. Despite individuals having been vaccinated with no animal reproductive data being carried out, those that have been detrimentally effected are once again left to fight for justice.

An ambivalent relationship between science, the media, regulatory systems, and commerce once more creates an atmosphere of uncertainty. A parallel can be drawn by this situation and how Jesus described the leaven of the Pharisees. The persecution that Jesus endured for challenging corruption remains relevant to this day. Just as with leaven in the Bible, the quality of thought can improve society and allow it to prosper, whereas, corrupt thoughts will obstruct progress and lead to hardship.

Bibliography

Chapter 1 and 2

King James version. *The Holy Bible.* Collins, UK; 2011.

Peters T. *Science and theology: The new consonance.* Oxford, UK: Westview Press; 1998.

Solter D. Mammalian cloning: Advances and limitations. *Nat. Gen.* 2000; 1: 199–207.

Today's English Version. *The Good News Bible.* Collins, UK; 2007.

Chapters 3 and 4

Mortimer RK. Evolution and variation of the yeast (*Saccharomyces*) genome. *Gen. Res.* 2000; 10: 403–409.

Phaff HJ, Miller MW, Mrak EM. *The life of yeasts.* Cambridge, MA, USA: Harvard University Press; 1968.

Rose AH, Harrison JS, Wheals AE. *The Yeasts*, London, UK: Academic Press; 1991.

Hall MN, Linder P. *The Early days of Yeast Genetics.* New York, USA: Cold Spring Harbour Laboratory Press; 1993.

Wilson EO. Consilience: The unity of knowledge. London, UK: Little, Brown and Co; 1998.

Chapter 5

Hoyte HM. The plagues of Egypt: what killed the animals and the firstborn? *Med. J. Aust.* 1993; 159: 285–286.

Panagiotakopulu E. Pharaonic Egypt and the origins of plague. *J. Biogeogr.* 2004; 31: 269–275.

Trevisanato SI. Ancient Egyptian doctors and the nature of the biblical plagues. *Med. Hypotheses*, 2005; 65: 811–813.

Chapter 6

Anderson BW. *The living world of the Old Testament.*, London, UK: Longmans, Green and Co; 1967.

Kee HC, Young FW. *The living world of the New Testament.* NJ, USA: Prentice-Hall Inc; 1960.

Chapter 7

Lockyer H. *All the parables of the Bible*. MI, USA: Zondervan;1988.

Chapter 8

Bacchiocchi S. *Wine in the bible: A biblical study on the use of alcoholic beverages*. MI, USA: Biblical Perspectives; 1989.

Parsons B. *The wine question settled: In accordance with the inductions of science and the facts of history*. London, UK: John Snow; 1841.

Chapter 9

Dujon B, Sherman D, Fischer G, et al. Genome evolution in yeasts. *Nature*. 2004; 430: 35–43.

Gasch AP. Comparative genomics of the environmental stress response in ascomycete fungi. *Yeast*. 2007; 24: 961–976.

Murray A, Hunt T. *The Cell Cycle. An Introduction*. New York and Oxford: Oxford University Press, UK; 1993.

Oliver SG. From DNA sequence to biological function. *Nature*. 1996; 379: 597–600.

Sambrook J, Fritsch EF, Mauriatis T. Molecular Cloning: A laboratory Manual. NY, USA: Cold Spring Harbour Laboratory Press; 1989.

Chapter 10

Araneta MR, Schlangen KM, Edmonds LD, et al. (2003). Prevalence of birth defects among infants of Gulf War veterans in Arkansas, Arizona, California, Bioport Corporation. Anthrax vaccine data sheet. Lansing, MI, USA US license No. 1260; Jan 2002.

Georgia, Hawaii, and Iowa, 1989-1993. *Birth Defects Res A Clin Mol Teratol.* 2003; 267: 246–260.

Goldacre B. *Bad science*. London, UK: Harper Perennial; 2009.

The Sunday Times Insight Team. *Suffer the children: The story of Thalidomide*. London, UK: Futura Publications; 1980.

About the author:

I have always been fascinated by the relationship between science and religion, so it seemed natural to write a popular science book about this subject. My specialist subject, and the focus of my doctorate thesis, was yeast genetics. This book represents how a molecular biologist has interpreted the microscopic world as viewed in the Biblical era. My aim was to demonstrate the impact the Bible has had on science using yeast for illustrative purposes. Hopefully, these views have been presented impartially, presenting the negatives and positives of both science and religion. While researching the Bible, I became intrigued at how humanity copes with uncertainties, and the role of religion and science in dealing with this. My hope is that this book will present the reader with a portrayal of how discovery, through science, influences social and religious views by eliminating some uncertainties.

www.ingramcontent.com/pod-product-compliance
Lightning Source LLC
Chambersburg PA
CBHW070852180526
45168CB00005B/1784